# MANIFESTLY HARAWAY

# THE CYBORG MANIFESTO

# THE COMPANION
# SPECIES MANIFESTO

# COMPANIONS IN
# CONVERSATION
# (WITH CARY WOLFE)

**posthumanities 37**

DONNA J. HARAWAY

# *Manifestly*
# *Haraway*

UNIVERSITY OF

MINNESOTA PRESS

MINNEAPOLIS

LONDON

Published by the University of Minnesota Press
111 Third Avenue South, Suite 290
Minneapolis, MN 55401-2520
http://www.upress.umn.edu

ISBN 978-0-8166-5047-7 (hc)
ISBN 978-0-8166-5048-4 (pb)

A Cataloging-in-Publication record for this book is available from the Library of Congress.

Printed in the United States of America on acid-free paper

The University of Minnesota is an equal-opportunity educator and employer.

27  26  25  24  23                10  9  8  7  6

# CONTENTS

# INTRODUCTION

In the thirty-plus years I've been reading critical and cultural
theory, I don't think there's ever been a phenomenon like "The
Cyborg Manifesto." I remember distinctly the first time I read
it (in the form of a dog-eared Xerox copy, as was the custom
among graduate students in those days). I've met lots of people
over the years who had the same experience with the manifesto—less like remembering where you were on 9/11 than re-
calling the first time you listened to a record that really blew you
away. On intellectual grounds, I was drawn to the text in part because as an undergraduate I had already become interested in
systems theory (or what was then often called "cybernetics"),
thanks in no small part to the work of Gregory Bateson in *Steps
to an Ecology of Mind*. (Only later—much later—would I discover the happy coincidence that both Bateson and Haraway
had taught and made their homes in Santa Cruz, California, in
the thick of what would become the History of Consciousness
Board and, later, Department.) I was prepared, then (at least in
part), for the interdisciplinary intellectual sweep of the manifesto and its mash-up of science, technoculture, science fiction, philosophy, socialist-feminist politics, and theory. But
what I wasn't prepared for—and I don't think many people
were—was its stylistic and rhetorical bravado, what I'd even call

its swagger (being deliberately heretical here, precisely in the spirit of the manifesto itself). Who else launches an essay with observations such as "Cyborg 'sex' restores some of the lovely replicative baroque of ferns and invertebrates (such nice organic prophylactics against heterosexism)" and ends it with the declaration "Though both are bound in the spiral dance, I'd rather be a cyborg than a goddess"?

It wasn't just that the manifesto made clear to me, in theoretical terms, something I would try to articulate later in my own work: that rethinking the so-called "question of the animal" was really a subset of a much broader challenge that would come to be called *posthumanism* (a term Haraway chafes against, for reasons we discuss in these pages). That much is announced barely three pages into the text, where we find the famous passage on the "three crucial boundary breakdowns" (between human and animal, organism and machine, and the physical and the nonphysical) that provides the point of entry for the manifesto to do its work. No, for me, and I'll wager for most readers, it was the unprecedented writerly whirlwind of the text that made it unforgettable—its swervings and foldings, the mix of tones, voices, and conjurings, winking at the reader here only to do some serious cage rattling on the very next page. A term that comes up a lot (both in our conversation here and in Haraway's own characterization of the text) is *irony,* but *irony* doesn't begin to capture the amazing range of tones, personae, and voices that Haraway is able to inhabit in these pages.

*Introduction*

The rhetorical performance is so stunning that it's easy to forget just how encyclopedic the text is, and how generous, too. How much do you first have to know to even contemplate a piece of writing such as this? And where else—in an era of academic stardom that was already well under way at the time—do we find a more generous citational practice (something Haraway takes very seriously, as readers of our conversation will discover)? Try making a list of just the proper names mentioned in the text. For these reasons (and more, of course), "The Cyborg Manifesto" was a profoundly liberating experience for many readers—not liberating as in "freedom to do whatever you like," but liberating in the sense of modeling for us a new and unprecedented range of expression and experimentation for serious academic writing. Given its headlong pace and its weave of affective registers and discursive textures, it sometimes felt more like reading a novel or experimental fiction than reading an academic essay. I think many readers left their first encounter with the manifesto thinking to themselves, "Wow, you can really write this way?!" Well, yes and no. You can if you're Donna Haraway.

But "The Cyborg Manifesto" is also very much a product of its moment, and this is as it should be, since cyborgs (as she reminds us many times in the text) have no truck with timelessness or immortality. Reading it again today, it's a sort of time capsule or cultural brain smear from the era of Star Wars (both the Hollywood film franchise and the Reagan-era missile

*Introduction*

defense system) blasphemously reinterpreted by a committed socialist-feminist who is ready to do something about it, is looking for help from you and me, and will use any and every tool in the shed to make a good start on the job. Almost twenty years later, Haraway had decided that the appropriate and necessary tools had changed, in part because of a very long and very serious involvement with dogs and dog training that first brought us together as friends (and brought us together as two people who felt that they could, partly on those grounds, understand and admire the late Vicki Hearne in ways that few officially functioning academics could). As she writes in "The Companion Species Manifesto," "I have come to see cyborgs as junior siblings in the much bigger, queer family of companion species."

Of course, the bio and the techno have always been completely intertwined in Haraway's work, early and late; they are wound up in that "spiral dance" that ends Manifesto I. But "The Companion Species Manifesto" reaches—and even yearns—toward that other pole of the bio/techno problematic, the flesh (keeping in mind, as she reminds us early in Manifesto II, that "these figures are hardly polar opposites"). While it's true, as Haraway writes in Manifesto II, that "neither a cyborg nor a companion animal pleases the pure of heart who long for better protected species boundaries and sterilization of category deviants," there's a need for flesh and earth here that gives the second manifesto a different feel. It's not just that these are revealed to be the site of a more densely woven complexity—ontologically, ethically, and politically—than the circuit, the chip, or the algorithm. It's also that this is less a story about technoscience (though it's obviously that, too) than a story of "biopower and biosociality," of how "history matters in naturecultures," including Haraway's own history (see "Notes of a Sports Writer's Daughter"), and including the history of this complicated creature called "The Australian Shepherd," and how *those* two end up in their own kind of "spiral dance," one that is less about cyborgs and goddesses than about bitches, messmates, and what Haraway (doing a number on Foucault) calls "the birth of the kennel."

To delve into the complexities—historical, genetic, and otherwise—of the AKC-recognized "purebred" dog is to enter fully biopolitical territory, because, as we now know, race is absolutely central to the work of biopolitics, and it's impossible to talk about race without talking about species. In light of all the biopolitical literature devoted to the Holocaust and the Nazi camps, the word *purebred* takes on a rather more ominous cast by the time the second manifesto appears. That text reminds us of something we ought not need reminding of: that Donna Haraway is one of the most important thinkers in the history of what is now officially called "biopolitical thought" (a genealogy that can often seem, with its procession of white, male, European continental philosophers, a little too purebred for its own good—though I am happy to see Roberto Esposito, a fellow *Posthumanities* author, giving Haraway her due). Does any

*Introduction*

major thinker of biopolitics (and I include Michel Foucault in that ascription) bring to the table anything like the range of disciplinary expertise and nimbleness across literatures, discourses, and political communities that we find in Haraway's writing? I don't think so. No doubt "The Companion Species Manifesto" (and Haraway's book *When Species Meet*, which grew out of it) makes a unique and remarkably timely contribution to biopolitical thought. But it also reminds us of something I was keen to underscore in these pages: that in essays such as "The Biological Enterprise: Sex, Mind, and Profit from Human Engineering to Sociobiology" (published in 1979), Haraway was "doing biopolitics" long before it became codified as a field.

Finally, to return to the title of this collection, readers will find late in our conversation the amplification of an often-overlooked element in Haraway's writing, one that will have a salutary effect, I hope, on what is often called "the return of religion" in contemporary theory and philosophy. After all, a primary sense of the word *manifest*, according to the Oxford English Dictionary, is "to make something evident to the eye or to the understanding . . . especially of supernatural beings," as in the example offered from *The Gospel of Ramakrishna*: "I see that God is walking in every human form and manifesting Himself alike through the sage and the sinner." As we discuss in some detail in our exchange, the powerful trope of "the word made flesh" does important work in Haraway's writing—not just conceptually, but ethically, politically, indeed *bio*politically—and it provides an important counterlogic for her to a certain binding, hegemonic matrix of secularism, Protestantism, capitalism, and the state form in the history of the modern United States. As she reveals, her attraction to this trope is born, in no small part, of the fact of being raised Catholic. But her own transubstantiation of it is due to the fact of being raised not just Catholic but a Catholic *woman* who comes of age and is educated in the era of Sputnik and the Space Race. To return full circle to the opening of "The Cyborg Manifesto," then, Haraway's use of "the word made flesh" is "blasphemy," perhaps, but all the more serious and all the more faithful because of it. After all, as she reminds us in the very letter of her work, "blasphemy is not apostasy."

# THE MANIFESTOS

# A Cyborg Manifesto

SCIENCE, TECHNOLOGY,

AND SOCIALIST-FEMINISM

IN THE LATE

TWENTIETH CENTURY

## AN IRONIC DREAM OF A
## COMMON LANGUAGE FOR WOMEN
## IN THE INTEGRATED CIRCUIT

This essay is an effort to build an ironic political myth faithful to feminism, socialism, and materialism. Perhaps more faithful as blasphemy is faithful, than as reverent worship and identification. Blasphemy has always seemed to require taking things very seriously. I know no better stance to adopt from within the secular-religious, evangelical traditions of United States politics, including the politics of socialist-feminism. Blasphemy protects one from the moral majority within, while still insisting on the need for community. Blasphemy is not apostasy. Irony is about contradictions that do not resolve into larger wholes, even dialectically, about the tension of holding incompatible things together because both or all are necessary and true. Irony is about humor and serious play. It is also a rhetorical strategy and a political method, one I would like to see more honored within socialist-feminism. At the center of my ironic faith, my blasphemy, is the image of the cyborg.

A cyborg is a cybernetic organism, a hybrid of machine and organism, a creature of social reality as well as a creature of fiction. Social reality is lived social relations, our most important political construction, a world-changing fiction. The international women's movements have constructed "women's experience," as well as uncovered or discovered this crucial collective object. This experience is a fiction and fact of the most crucial, political kind. Liberation rests on the construction of the consciousness, the imaginative apprehension, of oppression, and so of possibility. The cyborg is a matter of fiction and lived experience that changes what counts as women's experience in the late twentieth century. This is a struggle over life and death, but the boundary between science fiction and social reality is an optical illusion.

Contemporary science fiction is full of cyborgs — creatures simultaneously animal and machine, who populate worlds ambiguously natural and crafted. Modern medicine is also full of cyborgs, of couplings between organism and machine, each conceived as coded devices, in an intimacy and with a power that were not generated in the history of sexuality. Cyborg "sex" restores some of the lovely replicative baroque of ferns and invertebrates (such nice organic prophylactics against heterosexism). Cyborg replication is uncoupled from organic reproduction. Modem production seems like a dream of cyborg colonization work, a dream that makes the nightmare of Taylorism seem idyllic. And modern war is a cyborg orgy, coded by $C^3I$, command-control-communication-intelligence, an $84 billion item in 1984's U.S. defense budget. I am making an argument for the cyborg as a fiction mapping our social and bodily reality and as an imaginative resource suggesting some very fruitful couplings. Michel Foucault's biopolitics is a flaccid premonition of cyborg politics, a very open field.

By the late twentieth century, our time, a mythic time, we are all chimeras, theorized and fabricated hybrids of machine and organism—in short, cyborgs. The cyborg is our ontology; it gives us our politics. The cyborg is a condensed image of both imagination and material reality, the two joined centers structuring any possibility of historical transformation. In the traditions of "Western" science and politics—the tradition of racist, male-dominant capitalism; the tradition of progress; the tradition of the appropriation of nature as resource for the productions of culture; the tradition of reproduction of the self from the reflections of the other—the relation between organism and machine has been a border war. The stakes in the border war have been the territories of production, reproduction, and imagination. This essay is an argument for *pleasure* in the confusion of boundaries and for *responsibility* in their construction. It is also an effort to contribute to socialist-feminist culture and theory in a postmodernist, non-naturalist mode and in the utopian tradition of imagining a world without gender, which is perhaps a world without genesis, but maybe also a world without end. The cyborg incarnation is outside salvation history. Nor does it mark time on an oedipal calendar, attempting to heal the terrible cleavages of gender in an oral symbiotic utopia or post-oedipal apocalypse. As Zoë Sofoulis argues in her

*A Cyborg Manifesto*

unpublished manuscript on Jacques Lacan, Melanie Klein, and nuclear culture, "Lacklein," the most terrible and perhaps the most promising monsters in cyborg worlds are embodied in non-oedipal narratives with a different logic of repression, which we need to understand for our survival.[1]

The cyborg is a creature in a postgender world; it has no truck with bisexuality, pre-oedipal symbiosis, unalienated labor, or other seductions to organic wholeness through a final appropriation of all the powers of the parts into a higher unity. In a sense, the cyborg has no origin story in the Western sense—a "final" irony since the cyborg is also the awful apocalyptic *telos* of the "West's" escalating dominations of abstract individuation, an ultimate self untied at last from all dependency, a man in space. An origin story in the "Western," humanist sense depends on the myth of original unity, fullness, bliss and terror, represented by the phallic mother from whom all humans must separate, the task of individual development and of history, the twin potent myths inscribed most powerfully for us in psychoanalysis and Marxism. Hilary Klein has argued that both Marxism and psychoanalysis, in their concepts of labor and of individuation and gender formation, depend on the plot of original unity out of which difference must be produced and enlisted in a drama of escalating domination of woman/nature.[2] The cyborg skips the step of original unity, of identification with nature in the Western sense. This is its illegitimate promise that might lead to subversion of its teleology as Star Wars.

The cyborg is resolutely committed to partiality, irony, intimacy, and perversity. It is oppositional, utopian, and completely without innocence. No longer structured by the polarity of public and private, the cyborg defines a technological polis based partly on a revolution of social relations in the *oikos*, the household. Nature and culture are reworked; the one can no longer be the resource for appropriation or incorporation by the other. The relationships for forming wholes from parts, including those of polarity and hierarchical domination, are at issue in the cyborg world. Unlike the hopes of Frankenstein's monster, the cyborg does not expect its father to save it through a restoration of the garden—that is, through the fabrication of a heterosexual mate, through its completion in a finished whole, a city and cosmos. The cyborg does not dream of community on the model of the organic family, this time without the oedipal project. The cyborg would not recognize the Garden of Eden; it is not made of mud and cannot dream of returning to dust. Perhaps that is why I want to see if cyborgs can subvert the apocalypse of returning to nuclear dust in the manic compulsion to name the Enemy. Cyborgs are not reverent; they do not re-member the cosmos. They are wary of holism, but needy for connection—they seem to have a natural feel for united-front politics, but without the vanguard party. The main trouble with cyborgs, of course, is that they are the illegitimate offspring of militarism and patriarchal capitalism, not to mention state socialism. But illegitimate offspring are often

*A Cyborg Manifesto*

exceedingly unfaithful to their origins. Their fathers, after all, are inessential.

I will return to the science fiction of cyborgs at the end of this essay, but now I want to signal three crucial boundary break-downs that make the following political-fictional (political-scientific) analysis possible. By the late twentieth century in U.S. scientific culture, the boundary between human and animal is thoroughly breached. The last beachheads of uniqueness have been polluted if not turned into amusement parks: language, tool use, social behavior, mental events—nothing really convincingly settles the separation of human and animal. And many people no longer feel the need for such a separation; indeed, many branches of feminist culture affirm the pleasure of connection of human and other living creatures. Movements for animal rights are not irrational denials of human uniqueness; they are a clear-sighted recognition of connection across the discredited breach of nature and culture. Biology and evolutionary theory over the past two centuries have simultaneously produced modern organisms as objects of knowledge and reduced the line between humans and animals to a faint trace re-etched in ideological struggle or professional disputes between life and social science. Within this framework, teaching modern Christian creationism should be fought as a form of child abuse.

Biological-determinist ideology is only one position opened up in scientific culture for arguing the meanings of human animality. There is much room for radical political people to contest the meanings of the breached boundary.[3] The cyborg appears in myth precisely where the boundary between human and animal is transgressed. Far from signaling a walling off of people from other living beings, cyborgs signal disturbingly and pleasurably tight coupling. Bestiality has a new status in this cycle of marriage exchange.

The second leaky distinction is between animal–human (organism) and machine. Pre-cybernetic machines could be haunted; there was always the specter of the ghost in the machine. This dualism structured the dialogue between materialism and idealism that was settled by a dialectical progeny, called spirit or history, according to taste. But basically machines were not self-moving, self-designing, autonomous. They could not achieve man's dream, only mock it. They were not man, an author to himself, but only a caricature of that masculinist reproductive dream. To think they were otherwise was paranoid. Now we are not so sure. Late twentieth-century machines have made thoroughly ambiguous the difference between natural and artificial, mind and body, self-developing and externally designed, and many other distinctions that used to apply to organisms and machines. Our machines are disturbingly lively, and we ourselves frighteningly inert.

Technological determination is only one ideological space opened up by the reconceptions of machine and organism as coded texts through which we engage in the play of writing and reading the world.[4] "Textualization" of everything in post-structuralist, postmodernist theory has been damned by Marxists and socialist-feminists for its utopian disregard for the lived relations of domination that ground the "play" of arbitrary reading.[5] It is certainly true that postmodernist strategies, like my cyborg myth, subvert myriad organic wholes (for example, the poem, the primitive culture, the biological organism). In short, the certainty of what counts as nature—a source of insight and promise of innocence—is undermined, probably fatally. The transcendent authorization of interpretation is lost, and with it the ontology grounding "Western" epistemology. But the alternative is not cynicism or faithlessness, that is, some version of abstract existence, like the accounts of technological determinism destroying "man" by the "machine" or "meaningful political action" by the "text." Who cyborgs will be is a radical question; the answers are a matter of survival. Both chimpanzees and artifacts have politics (de Waal 1982; Winner 1980), so why shouldn't we?

The third distinction is a subset of the second: the boundary between physical and nonphysical is very imprecise for us. Pop physics books on the consequences of quantum theory and the indeterminacy principle are a kind of popular scientific equivalent to Harlequin romances[6] as a marker of radical change in American white heterosexuality: they get it wrong, but they are on the right subject. Modern machines are quintessentially microelectronic devices: they are everywhere and they are invisible. Modern machinery is an irreverent upstart god, mocking the Father's ubiquity and spirituality. The silicon chip is a surface for writing; it is etched in molecular scales disturbed only by atomic noise, the ultimate interference for nuclear scores.

Writing, power, and technology are old partners in Western stories of the origin of civilization, but miniaturization has changed our experience of mechanism. Miniaturization has turned out to be about power; small is not so much beautiful as preeminently dangerous, as in cruise missiles. Contrast the TV sets of the 1950s or the news cameras of the 1970s with the TV wristbands or hand-sized video cameras now advertised. Our best machines are made of sunshine; they are all light and clean because they are nothing but signals, electromagnetic waves, a section of a spectrum, and these machines are eminently portable, mobile—a matter of immense human pain in Detroit and Singapore. People are nowhere near so fluid, being both material and opaque. Cyborgs are ether, quintessence.

The ubiquity and invisibility of cyborgs are precisely why these Sunshine Belt machines are so deadly. They are as hard to see politically as materially. They are about consciousness—or its simulation.[7] They are floating signifiers moving in pickup trucks across Europe, blocked more effectively by the witch-weavings of the displaced and so unnatural women of the anti-nuclear Greenham Women's Peace Camp, who read the cyborg webs of power so very well, than by the militant labor of older masculinist politics, whose natural constituency needs defense

*A Cyborg Manifesto*

jobs. Ultimately the "hardest" science is about the realm of greatest boundary confusion, the realm of pure number, pure spirit, $C^3I$, cryptography, and the preservation of potent secrets. The new machines are so clean and light. Their engineers are sun-worshippers mediating a new scientific revolution associated with the night dream of postindustrial society. The diseases evoked by these clean machines are "no more" than the minuscule coding changes of an antigen in the immune system, "no more" than the experience of stress. The nimble fingers of "Oriental" women, the old fascination of little Anglo-Saxon Victorian girls with doll's houses, women's enforced attention to the small take on quite new dimensions in this world. There might be a cyborg Alice taking account of these new dimensions. Ironically, it might be the unnatural cyborg women making chips in Asia and spiral dancing in Santa Rita Jail[8] whose constructed unities will guide effective oppositional strategies.

So my cyborg myth is about transgressed boundaries, potent fusions, and dangerous possibilities, which progressive people might explore as one part of needed political work. One of my premises is that most American socialists and feminists see deepened dualisms of mind and body, animal and machine, idealism and materialism in the social practices, symbolic formulations, and physical artifacts associated with "high technology" and scientific culture. From *One-Dimensional Man* (Marcuse 1964) to *The Death of Nature* (Merchant 1980), the analytic resources developed by progressives have insisted on the necessary domination of technics and recalled us to an imagined organic body to integrate our resistance. Another of my premises is that the need for unity of people trying to resist worldwide intensification of domination has never been more acute. But a slightly perverse shift of perspective might better enable us to contest for meanings, as well as for other forms of power and pleasure in technologically mediated societies.

From one perspective, a cyborg world is about the final imposition of a grid of control on the planet, about the final abstraction embodied in a Star Wars apocalypse waged in the name of defense, about the final appropriation of women's bodies in a masculinist orgy of war (Sofia 1984). From another perspective, a cyborg world might be about lived social and bodily realities in which people are not afraid of their joint kinship with animals and machines, not afraid of permanently partial identities and contradictory standpoints. The political struggle is to see from both perspectives at once because each reveals both dominations and possibilities unimaginable from the other vantage point. Single vision produces worse illusions than double vision or many-headed monsters. Cyborg unities are monstrous and illegitimate; in our present political circumstances, we could hardly hope for more potent myths for resistance and recoupling. I like to imagine LAG, the Livermore Action Group, as a kind of cyborg society, dedicated to realistically converting the laboratories that most fiercely embody and spew out the

*A Cyborg Manifesto*

15

tools of technological apocalypse, and committed to building a political form that actually manages to hold together witches, engineers, elders, perverts, Christians, mothers, and Leninists long enough to disarm the state. Fission Impossible is the name of the affinity group in my town. (Affinity: related not by blood but by choice, the appeal of one chemical nuclear group for another, avidity.)[9]

## FRACTURED IDENTITIES

It has become difficult to name one's feminism by a single adjective—or even to insist in every circumstance on the noun. Consciousness of exclusion through naming is acute. Identities seem contradictory, partial, and strategic. With the hard-won recognition of their social and historical constitution, gender, race, and class cannot provide the basis for belief in "essential" unity. There is nothing about being "female" that naturally binds women. There is not even such a state as "being" female, itself a highly complex category constructed in contested sexual scientific discourses and other social practices. Gender, race, or class consciousness is an achievement forced on us by the terrible historical experience of the contradictory social realities of patriarchy, colonialism, and capitalism. And who counts as "us" in my own rhetoric? Which identities are available to ground such a potent political myth called "us," and what could motivate enlistment in this collectivity? Painful fragmentation among feminists (not to mention among women) along every possible fault line has made the concept of *woman* elusive, an excuse for the matrix of women's dominations of each other. For me—and for many who share a similar historical location in white, professional middle-class, female, radical, North American, mid-adult bodies—the sources of a crisis in political identity are legion. The recent history for much of the U.S. left and U.S. feminism has been a response to this kind of crisis by endless splitting and searches for a new essential unity. But there has also been a growing recognition of another response through coalition—affinity, not identity.[10]

Chela Sandoval (n.d.; 1984), from a consideration of specific historical moments in the formation of the new political voice called women of color, has theorized a hopeful model of political identity called "oppositional consciousness," born of the skills for reading webs of power by those refused stable membership in the social categories of race, sex, or class. *Women of color,* a name contested at its origins by those whom it would incorporate, as well as a historical consciousness marking systematic breakdown of all the signs of Man in "Western" traditions, constructs a kind of postmodernist identity out of otherness, difference, and specificity. This postmodernist identity is fully political, whatever might be said about other possible postmodernisms. Sandoval's oppositional consciousness is about contradictory locations and heterochronic calendars, not about relativisms and pluralisms.

Sandoval emphasizes the lack of any essential criterion for identifying who is a woman of color. She notes that the definition of the group has been by conscious appropriation of negation. For example, a Chicana or U.S. black woman has not been able to speak as a woman or as a black person or as a Chicano. Thus, she was at the bottom of a cascade of negative identities, left out of even the privileged oppressed authorial categories called "women and blacks," who claimed to make the important revolutions. The category "woman" negated all nonwhite women; "black" negated all nonblack people, as well as all black women. But there was also no "she," no singularity, but a sea of differences among U.S. women who have affirmed their historical identity as U.S. women of color. This identity marks out a self-consciously constructed space that cannot affirm the capacity to act on the basis of natural identification, but only on the basis of conscious coalition, of affinity, of political kinship.[11] Unlike the "woman" of some streams of the white women's movement in the United States, there is no naturalization of the matrix, or at least this is what Sandoval argues is uniquely available through the power of oppositional consciousness.

Sandoval's argument has to be seen as one potent formulation for feminists out of the worldwide development of anticolonialist discourse; that is to say, discourse dissolving the

"West" and its highest product—the one who is not animal, barbarian, or woman; man, that is, the author of a cosmos called history. As orientalism is deconstructed politically and semiotically, the identities of the occident destabilize, including those of feminists.[12] Sandoval argues that "women of color" have a chance to build an effective unity that does not replicate the imperializing, totalizing revolutionary subjects of previous Marxisms and feminisms, which had not faced the consequences of the disorderly polyphony emerging from decolonization.

Katie King has emphasized the limits of identification and the political/poetic mechanics of identification built into reading "the poem," that generative core of cultural feminism. King criticizes the persistent tendency among contemporary feminists from different "moments" or "conversations" in feminist practice to taxonomize the women's movement to make one's own political tendencies appear to be the *telos* of the whole. These taxonomies tend to remake feminist history so that it appears to be an ideological struggle among coherent types persisting over time, especially those typical units called radical, liberal, and socialist-feminism. Literally, all other feminisms are either incorporated or marginalized, usually by building an explicit ontology and epistemology.[13] Taxonomies of feminism produce epistemologies to police deviation from official women's experience. And of course, "women's culture," like women of color, is consciously created by mechanisms induc-

*A Cyborg Manifesto*

19

ing affinity. The rituals of poetry, music, and certain forms of academic practice have been preeminent. The politics of race and culture in the U.S. women's movements are intimately interwoven. The common achievement of King and Sandoval is learning how to craft a poetic/political unity without relying on a logic of appropriation, incorporation, and taxonomic identification.

The theoretical and practical struggle against unity-through-domination or unity-through-incorporation ironically undermines not only the justifications for patriarchy, colonialism, humanism, positivism, essentialism, scientism, and other unlamented -isms, but *all* claims for an organic or natural standpoint. I think that radical and socialist/Marxist-feminisms have also undermined their/our own epistemological strategies and that this is a crucially valuable step in imagining possible unities. It remains to be seen whether all "epistemologies" as Western political people have known them fail us in the task to build effective affinities.

It is important to note that the effort to construct revolutionary standpoints, epistemologies as achievements of people committed to changing the world, has been part of the process showing the limits of identification. The acid tools of postmodernist theory and the constructive tools of ontological discourse about revolutionary subjects might be seen as ironic allies in dissolving Western selves in the interests of survival. We are excruciatingly conscious of what it means to have a histor-

*A Cyborg Manifesto*

ically constituted body. But with the loss of innocence in our origin, there is no expulsion from the Garden either. Our politics lose the indulgence of guilt with the *naïveté* of innocence. But what would another political myth for socialist-feminism look like? What kind of politics could embrace partial, contradictory, permanently unclosed constructions of personal and collective selves and still be faithful, effective—and, ironically, socialist-feminist?

I do not know of any other time in history when there was greater need for political unity to confront effectively the dominations of "race," "gender," "sexuality," and "class." I also do not know of any other time when the kind of unity we might help build could have been possible. None of "us" has any longer the symbolic or material capability of dictating the shape of reality to any of "them." Or at least "we" cannot claim innocence from practicing such dominations. White women, including socialist-feminists, discovered (that is, were forced kicking and screaming to notice) the noninnocence of the category "woman." That consciousness changes the geography of all previous categories; it denatures them as heat denatures a fragile protein. Cyborg feminists have to argue that "we" do not want any more natural matrix of unity and that no construction is whole. Innocence, and the corollary insistence on victimhood as the only ground for insight, has done enough damage. But the constructed revolutionary subject must give late-twentieth-century people pause as well. In the fraying of identities and in the reflexive strategies for constructing them, the possibility opens up for weaving something other than a shroud for the day after the apocalypse that so prophetically ends salvation history.

Both Marxist/socialist-feminisms and radical feminisms have simultaneously naturalized and denatured the category "woman" and consciousness of the social lives of "women." Perhaps a schematic caricature can highlight both kinds of moves. Marxian socialism is rooted in an analysis of wage labor that reveals class structure. The consequence of the wage relationship is systematic alienation, as the worker is dissociated from his (*sic*) product. Abstraction and illusion rule in knowledge, domination rules in practice. Labor is the preeminently privileged category enabling the Marxist to overcome illusion and find that point of view that is necessary for changing the world. Labor is the humanizing activity that makes man; labor is an ontological category permitting the knowledge of a subject, and so the knowledge of subjugation and alienation.

In faithful filiation, socialist-feminism advanced by allying itself with the basic analytic strategies of Marxism. The main achievement of both Marxist feminists and socialist feminists was to expand the category of labor to accommodate what (some) women did, even when the wage relation was subordinated to a more comprehensive view of labor under capitalist patriarchy. In particular, women's labor in the household and women's activity as mothers generally (that is, reproduction in the socialist-feminist sense) entered theory on the authority of analogy to the Marxian concept of labor. The unity of women here rests on an epistemology based on the ontological structure of "labor." Marxist/socialist-feminism does not "naturalize" unity; it is a possible achievement based on a possible standpoint rooted in social relations. The essentializing move is in the ontological structure of labor or of its analogue, women's activity.[14] The inheritance of Marxian humanism, with its preeminently Western self, is the difficulty for me. The contribution from these formulations has been the emphasis on the daily responsibility of real women to build unities, rather than to naturalize them.

Catharine MacKinnon's (1982, 1987) version of radical feminism is itself a caricature of the appropriating, incorporating, totalizing tendencies of Western theories of identity grounding action.[15] It is factually and politically wrong to assimilate all of the diverse "moments" or "conversations" in recent women's politics named radical feminism to MacKinnon's version. But the teleological logic of her theory shows how an epistemology and ontology—including their negations—erase or police difference. Only one of the effects of MacKinnon's theory is the rewriting of the history of the polymorphous field called radical feminism. The major effect is the production of a theory of experience, of women's identity, that is a kind of apocalypse for all revolutionary standpoints. That is, the totalization built into this tale of radical feminism achieves its end—the unity of women—by enforcing the experience of and testimony to radical nonbeing. As for the Marxist/socialist-feminist, consciousness is an achievement, not a natural fact. And MacKinnon's theory eliminates some of the difficulties built into humanist revolutionary subjects, but at the cost of radical reductionism.

MacKinnon argues that feminism necessarily adopted a different analytical strategy from Marxism, looking first not at the structure of class but at the structure of sex/gender and its generative relationship, men's constitution and appropriation of women sexually. Ironically, MacKinnon's "ontology" constructs a nonsubject, a nonbeing. Another's desire, not the self's labor, is the origin of "woman." She therefore develops a theory of consciousness that enforces what can count as "women's" experience—anything that names sexual violation, indeed, sex itself as far as "women" can be concerned. Feminist practice is the construction of this form of consciousness—that is, the self-knowledge of a self-who-is-not.

Perversely, sexual appropriation in this feminism still has the epistemological status of labor; that is to say, the point from which an analysis able to contribute to changing the world must flow. But sexual objectification, not alienation, is the consequence of the structure of sex/gender. In the realm of knowledge, the result of sexual objectification is illusion and abstraction. However, a woman is not simply alienated from her product but in a deep sense does not exist as a subject, or even potential subject, since she owes her existence as a woman to sexual appropriation. To be constituted by another's desire is

*A Cyborg Manifesto*

not the same thing as to be alienated in the violent separation of the laborer from his product.

MacKinnon's radical theory of experience is totalizing in the extreme; it does not so much marginalize as obliterate the authority of any other women's political speech and action. It is a totalization producing what Western patriarchy itself never succeeded in doing—feminists' consciousness of the nonexistence of women, except as products of men's desire. I think MacKinnon correctly argues that no Marxian version of identity can firmly ground women's unity. But in solving the problem of the contradictions of any Western revolutionary subject for feminist purposes, she develops an even more authoritarian doctrine of experience. If my complaint about socialist/Marxian standpoints is their unintended erasure of polyvocal, unassimilable, radical difference made visible in anticolonial discourse and practice, MacKinnon's intentional erasure of all difference through the device of the "essential" nonexistence of women is not reassuring.

In my taxonomy, which like any other taxonomy is a reinscription of history, radical feminism can accommodate all the activities of women named by socialist-feminists as forms of labor only if the activity can somehow be sexualized. Reproduction had different tones of meanings for the two tendencies, one rooted in labor, one in sex, both calling the consequences of domination and ignorance of social and personal reality "false consciousness."

Beyond either the difficulties or the contributions in the argument of any one author, neither Marxist nor radical feminist points of view have tended to embrace the status of a partial explanation; both were regularly constituted as totalities. Western explanation has demanded as much; how else could the "Western" author incorporate its others? Each tried to annex other forms of domination by expanding its basic categories through analogy, simple listing, or addition. Embarrassed silence about race among white radical and socialist-feminists was one major, devastating political consequence. History and polyvocality disappear into political taxonomies that try to establish genealogies. There was no structural room for race (or for much else) in theory claiming to reveal the construction of the category woman and social group women as a unified or totalizable whole. The structure of my caricature looks like this:

socialist-feminism—structure of class // wage labor // alienation
labor, by analogy reproduction, by extension sex, by addition race

radical feminism—structure of gender // sexual appropriation // objectification
sex, by analogy labor, by extension reproduction, by addition race

In another context, the French theorist Julia Kristeva claimed that women appeared as a historical group after the Second

World War, along with groups like youth.[16] Her dates are doubtful; but we are now accustomed to remembering that as objects of knowledge and as historical actors "race" did not always exist, "class" has a historical genesis, and "homosexuals" are quite junior. It is no accident that the symbolic system of the family of man—and so the essence of woman—breaks up at the same moment that networks of connection among people on the planet are unprecedentedly multiple, pregnant, and complex. "Advanced capitalism" is inadequate to convey the structure of this historical moment. In the "Western" sense, the end of man is at stake. It is no accident that woman disintegrates into women in our time. Perhaps socialist-feminists were not substantially guilty of producing essentialist theory that suppressed women's particularity and contradictory interests. I think we have been, at least through unreflective participation in the logics, languages, and practices of white humanism and through searching for a single ground of domination to secure our revolutionary voice. Now we have less excuse. But in the consciousness of our failures, we risk lapsing into boundless difference and giving up on the confusing task of making partial, real connection. Some differences are playful; some are poles of world historical systems of domination. "Epistemology" is about knowing the difference.

*A Cyborg Manifesto*

In this attempt at an epistemological and political position, I would like to sketch a picture of possible unity, a picture indebted to socialist and feminist principles of design. The frame for my sketch is set by the extent and importance of rearrangements in worldwide social relations tied to science and technology. I argue for a politics rooted in claims about fundamental changes in the nature of class, race, and gender in an emerging system of world order analogous in its novelty and scope to that created by industrial capitalism; we are living through a movement from an organic, industrial society to a polymorphous, information system—from all work to all play, a deadly game. Simultaneously material and ideological, the dichotomies may be expressed in the following chart of transitions from the comfortable old hierarchical dominations to the scary new networks I have called the informatics of domination:

| Organics of Domination | Informatics of Domination |
|---|---|
| representation | > simulation |
| bourgeois novel, realism | > science fiction, postmodernism |
| organism | > biotic component |
| depth, integrity | > surface, boundary |
| heat | > noise |
| biology as clinical practice | > biology as inscription |

*A Cyborg Manifesto*

| | | |
|---|---|---|
| physiology | > | communications engineering |
| small group | > | subsystem |
| perfection | > | optimization |
| eugenics | > | population control |
| decadence, *Magic Mountain* | > | obsolescence, *Future Shock* |
| hygiene | > | stress management |
| microbiology, tuberculosis | > | immunology, AIDS |
| organic division of labor | > | ergonomics, cybernetics of labor |
| functional specialization | > | modular construction |
| reproduction | > | replication |
| organic sex role specialization | > | optimal genetic strategies |
| biological determinism | > | evolutionary inertia, constraints |
| community ecology | > | ecosystem |
| racial chain of being | > | neoimperialism, United Nations humanism |
| scientific management in home/factory | > | global factory/electronic cottage industry |
| family/market/factory | > | women in the integrated circuit |
| family wage | > | comparable worth |
| public/private | > | cyborg citizenship |
| nature/culture | > | fields of difference |
| cooperation | > | communications enhancement |

*A Cyborg Manifesto*

| | | |
|---|---|---|
| Freud | > | Lacan |
| sex | > | genetic engineering |
| labor | > | robotics |
| mind | > | artificial intelligence |
| World War II | > | Star Wars |
| white capitalist patriarchy | > | informatics of domination |

----

*Transitions from the comfortable old hierarchical dominations to the scary new networks of informatics of domination.*

This list suggests several interesting things.[17] First, the objects on the right-hand side cannot be coded as "natural," a realization that subverts naturalistic coding for the left-hand side as well. We cannot go back ideologically or materially. It's not just that "god" is dead; so is the "goddess." Or both are revivified in the worlds charged with microelectronic and biotechnological politics. In relation to objects like biotic components, one must think not in terms of essential properties, but in terms of design, boundary constraints, rates of flows, systems logics, costs of lowering constraints. Sexual reproduction is one kind of reproductive strategy among many, with costs and benefits as a function of the system environment. Ideologies of sexual reproduction can no longer reasonably call on notions of sex and sex role as organic aspects in natural objects like organisms and families. Such reasoning will be unmasked as irrational, and ironically corporate executives reading *Playboy* and antiporn

*A Cyborg Manifesto*

radical feminists will make strange bedfellows in jointly un-masking the irrationalism.

Likewise for race, ideologies about human diversity have to be formulated in terms of frequencies of parameters, like blood groups or intelligence scores. It is "irrational" to invoke concepts like primitive and civilized. For liberals and radicals, the search for integrated social systems gives way to a new practice called "experimental ethnography" in which an organic object dissipates in attention to the play of writing. At the level of ideology, we see translations of racism and colonialism into languages of development and underdevelopment, rates and constraints of modernization. Any objects or persons can be reasonably thought of in terms of disassembly and reassembly; no "natural" architectures constrain system design. The financial districts in all the world's cities, as well as the export-processing and free trade zones, proclaim this elementary fact of "late capitalism." The entire universe of objects that can be known scientifically must be formulated as problems in communications engineering (for the managers) or theories of the text (for those who would resist). Both are cyborg semiologies.

One should expect control strategies to concentrate on boundary conditions and interfaces, on rates of flow across boundaries—and not on the integrity of natural objects. "Integrity" or "sincerity" of the Western self gives way to decision procedures and expert systems. For example, control strategies applied to women's capacities to give birth to new human beings will be developed in the languages of population control and maximization of goal achievement for individual decision-makers. Control strategies will be formulated in terms of rates, costs of constraints, degrees of freedom. Human beings, like any other component or subsystem, must be localized in a system architecture whose basic modes of operation are probabilistic, statistical. No objects, spaces, or bodies are sacred in themselves; any component can be interfaced with any other if the proper standard, the proper code, can be constructed for processing signals in a common language. Exchange in this world transcends the universal translation effected by capitalist markets that Marx analyzed so well. The privileged pathology affecting all kinds of components in this universe is stress—communications breakdown (Hogness 1983). The cyborg is not subject to Foucault's biopolitics; the cyborg simulates politics, a much more potent field of operations.

This kind of analysis of scientific and cultural objects of knowledge that have appeared historically since the Second World War prepares us to notice some important inadequacies in feminist analysis that has proceeded as if the organic, hierarchical dualisms ordering discourse in "the West" since Aristotle still ruled. They have been cannibalized, or as Zoë Sofia (1984) might put it, they have been "techno-digested." The dichotomies between mind and body, animal and human, organism and machine, public and private, nature and culture, men and women, primitive and civilized are all in question ideologically.

The actual situation of women is their integration/exploitation into a world system of production/reproduction and communication called the informatics of domination. The home, workplace, market, public arena, the body itself — all can be dispersed and interfaced in nearly infinite, polymorphous ways, with large consequences for women and others — consequences that themselves are very different for different people and that make potent oppositional international movements difficult to imagine and essential for survival. One important route for reconstructing socialist-feminist politics is through theory and practice addressed to the social relations of science and technology, including crucially the systems of myth and meanings structuring our imaginations. The cyborg is a kind of disassembled and reassembled, postmodern collective and personal self. This is the self feminists must code.

Communications technologies and biotechnologies are the crucial tools recrafting our bodies. These tools embody and enforce new social relations for women worldwide. Technologies and scientific discourses can be partially understood as formalizations, i.e., as frozen moments, of the fluid social interactions constituting them, but they should also be viewed as instruments for enforcing meanings. The boundary is permeable between tool and myth, instrument and concept, historical systems of social relations and historical anatomies of possible bodies, including objects of knowledge. Indeed, myth and tool mutually constitute each other.

*A Cyborg Manifesto*

33

Furthermore, communications sciences and modern biologies are constructed by a common move—*the translation of the world into a problem of coding,* a search for a common language in which all resistance to instrumental control disappears and all heterogeneity can be submitted to disassembly, reassembly, investment, and exchange.

In communications sciences, the translation of the world into a problem in coding can be illustrated by looking at cybernetic (feedback-controlled) systems theories applied to telephone technology, computer design, weapons deployment, or database construction and maintenance. In each case, solution to the key questions rests on a theory of language and control; the key operation is determining the rates, directions, and probabilities of flow of a quantity called information. The world is subdivided by boundaries differentially permeable to information. Information is just that kind of quantifiable element (unit, basis of unity) that allows universal translation, and so unhindered instrumental power (called effective communication). The biggest threat to such power is interruption of communication. Any system breakdown is a function of stress. The fundamentals of this technology can be condensed into the metaphor $C^3I$, command-control-communication-intelligence, the military's symbol for its operations theory.

In modern biologies, the translation of the world into a problem in coding can be illustrated by molecular genetics, ecology, sociobiological evolutionary theory, and immunobiology. The organism has been translated into problems of genetic coding and readout. Biotechnology, a writing technology, informs research broadly.[18] In a sense, organisms have ceased to exist as objects of knowledge, giving way to biotic components, i.e., special kinds of information-processing devices. The analogous moves in ecology could be examined by probing the history and utility of the concept of the ecosystem. Immunobiology and associated medical practices are rich exemplars of the privilege of coding and recognition systems as objects of knowledge, as constructions of bodily reality for us. Biology here is a kind of cryptography. Research is necessarily a kind of intelligence activity. Ironies abound. A stressed system goes awry; its communication processes break down; it fails to recognize the difference between self and other. Human babies with baboon hearts evoke national ethical perplexity—for animal rights activists at least as much as for the guardians of human purity. In the United States gay men and intravenous drug users are the "privileged" victims of an awful immune system disease that marks (inscribes on the body) confusion of boundaries and moral pollution (Treichler 1987).

But these excursions into communications sciences and biology have been at a rarefied level; there is a mundane, largely economic reality to support my claim that these sciences and technologies indicate fundamental transformations in the structure of the world for us. Communications technologies depend on electronics. Modern states, multinational corporations,

*A Cyborg Manifesto*

military power, welfare state apparatuses, satellite systems, political processes, fabrication of our imaginations, labor-control systems, medical constructions of our bodies, commercial pornography, the international division of labor, and religious evangelism depend intimately on electronics. Microelectronics is the technical basis of simulacra—that is, of copies without originals.

Microelectronics mediates the translations of labor into robotics and word processing, sex into genetic engineering and reproductive technologies, and mind into artificial intelligence and decision procedures. The new biotechnologies concern more than human reproduction. Biology as a powerful engineering science for redesigning materials and processes has revolutionary implications for industry, perhaps most obvious today in areas of fermentation, agriculture, and energy. Communications sciences and biology are constructions of natural-technical objects of knowledge in which the difference between machine and organism is thoroughly blurred; mind, body, and tool are on very intimate terms. The "multinational" material organization of the production and reproduction of daily life and the symbolic organization of the production and reproduction of culture and imagination seem equally implicated. The boundary-maintaining images of base and superstructure, public and private, or material and ideal never seemed more feeble.

I have used Rachel Grossman's (1980) image of women in the

*A Cyborg Manifesto*

integrated circuit to name the situation of women in a world so intimately restructured through the social relations of science and technology.[19] I used the odd circumlocution *the social relations of science and technology* to indicate that we are not dealing with a technological determinism, but with a historical system depending on structured relations among people. But the phrase should also indicate that science and technology provide fresh sources of power, that we need fresh sources of analysis and political action (Latour 1984). Some of the rearrangements of race, sex, and class rooted in high-tech–facilitated social relations can make socialist-feminism more relevant to effective progressive politics.

## THE HOMEWORK ECONOMY "OUTSIDE THE HOME"

The "New Industrial Revolution" is producing a new worldwide working class, as well as new sexualities and ethnicities. The extreme mobility of capital and the emerging international division of labor are intertwined with the emergence of new collectivities, and the weakening of familiar groupings. These developments are neither gender- nor race-neutral. White men in advanced industrial societies have become newly vulnerable to permanent job loss, and women are not disappearing from the job rolls at the same rates as men. It is not simply that women in

*A Cyborg Manifesto*

Third World countries are the preferred labor force for the science-based multinationals in the export-processing sectors, particularly in electronics. The picture is more systematic and involves reproduction, sexuality, culture, consumption, and production. In the prototypical Silicon Valley, many women's lives have been structured around employment in electronics-dependent jobs, and their intimate realities include serial heterosexual monogamy, negotiating childcare, distance from extended kin or most other forms of traditional community, a high likelihood of loneliness and extreme economic vulnerability as they age. The ethnic and racial diversity of women in Silicon Valley structures a microcosm of conflicting differences in culture, family, religion, education, and language.

Richard Gordon has called this new situation the "homework economy."[20] Although he includes the phenomenon of literal homework emerging in connection with electronics assembly, Gordon intends *homework economy* to name a restructuring of work that broadly has the characteristics formerly ascribed to female jobs, jobs literally done only by women. Work is being redefined as both literally female and feminized, whether performed by men or women. To be feminized means to be made extremely vulnerable; able to be disassembled, reassembled, exploited as a reserve labor force; seen less as workers than as servers; subjected to time arrangements on and off the paid job that make a mockery of a limited workday; leading an existence that always borders on being obscene, out of place, and reducible to sex. Deskilling is an old strategy newly applicable to formerly privileged workers. However, the homework economy does not refer only to large-scale deskilling, nor does it deny that new areas of high skill are emerging, even for women and men previously excluded from skilled employment. Rather, the concept indicates that factory, home, and market are integrated on a new scale and that the places of women are crucial— and need to be analyzed for differences among women and for meanings for relations between men and women in various situations.

The homework economy as a world capitalist organizational structure is made possible by (not caused by) the new technologies. The success of the attack on relatively privileged, mostly white, men's unionized jobs is tied to the power of the new communications technologies to integrate and control labor despite extensive dispersion and decentralization. The consequences of the new technologies are felt by women both in the loss of the family (male) wage (if they ever had access to this white privilege) and in the character of their own jobs, which are becoming capital-intensive—for example, office work and nursing.

The new economic and technological arrangements are also related to the collapsing welfare state and the ensuing intensification of demands on women to sustain daily life for themselves as well as for men, children, and old people. The feminization of poverty—generated by dismantling the welfare state, by the homework economy where stable jobs become the exception,

*A Cyborg Manifesto*

and sustained by the expectation that women's wages will not be matched by a male income for the support of children—has become an urgent focus. The causes of various women-headed households are a function of race, class, or sexuality; but their increasing generality is a ground for coalitions of women on many issues. That women regularly sustain daily life partly as a function of their enforced status as mothers is hardly new; the kind of integration with the overall capitalist and progressively war-based economy is new. The particular pressure, for example, on U.S. black women, who have achieved an escape from (barely) paid domestic service and who now hold clerical and similar jobs in large numbers, has large implications for continued enforced black poverty with employment. Teenage women in industrializing areas of the Third World increasingly find themselves the sole or major source of a cash wage for their families, while access to land is ever more problematic. These developments must have major consequences in the psychodynamics and politics of gender and race.

Within the framework of three major stages of capitalism (commercial/early industrial, monopoly, multinational)—tied to nationalism, imperialism, and multinationalism, and related to Jameson's three dominant aesthetic periods of realism, modernism, and postmodernism—I would argue that specific forms of families dialectically relate to forms of capital and to its political and cultural concomitants. Although lived problematically and unequally, ideal forms of these families might be schematized as (1) the patriarchal nuclear family, structured by the dichotomy between public and private and accompanied by the white bourgeois ideology of separate spheres and nineteenth-century Anglo-American bourgeois feminism; (2) the modern family mediated (or enforced) by the welfare state and institutions like the family wage, with a flowering of a-feminist heterosexual ideologies, including their radical versions represented in Greenwich Village around the First World War; and (3) the "family" of the homework economy with its oxymoronic structure of women-headed households and its explosion of feminisms and the paradoxical intensification and erosion of gender itself.

This is the context in which the projections for worldwide structural unemployment stemming from the new technologies are part of the picture of the homework economy. As robotics and related technologies put men out of work in "developed" countries and exacerbate failure to generate male jobs in Third World "development," and as the automated office becomes the rule even in labor-surplus countries, the feminization of work intensifies. Black women in the United States have long known what it looks like to face the structural underemployment ("feminization") of black men, as well as their own highly vulnerable position in the wage economy. It is no longer a secret that sexuality, reproduction, family, and community life are interwoven with this economic structure in myriad ways that have also differentiated the situations of white and black

*A Cyborg Manifesto*

women. Many more women and men will contend with similar situations, which will make cross-gender and race alliances on issues of basic life support (with or without jobs) necessary, not just nice.

The new technologies also have a profound effect on hunger and on food production for subsistence worldwide. Rae Lessor Blumberg (1981) estimates that women produce about 50 percent of the world's subsistence food.[21] Women are excluded generally from benefiting from the increased high-tech commodification of food and energy crops, their days are made more arduous because their responsibilities to provide food do not diminish, and their reproductive situations are made more complex. Green Revolution technologies interact with other high-tech industrial production to alter gender divisions of labor and differential gender migration patterns.

The new technologies seem deeply involved in the forms of "privatization" that Rosalind Petchesky (1981) has analyzed, in which militarization, right-wing family ideologies and policies, and intensified definitions of corporate (and state) property as private synergistically interact.[22] The new communications technologies are fundamental to the eradication of "public life" for everyone. This facilitates the mushrooming of a permanent high-tech military establishment at the cultural and economic expense of most people, but especially of women. Technologies like video games and highly miniaturized televisions seem crucial to production of modern forms of "private life." The culture

*A Cyborg Manifesto*

of video games is heavily oriented to individual competition and extraterrestrial warfare. High-tech, gendered imaginations are produced here, imaginations that can contemplate destruction of the planet and a sci-fi escape from its consequences. More than our imaginations is militarized; and the other realities of electronic and nuclear warfare are inescapable. These are the technologies that promise ultimate mobility and perfect exchange—and incidentally enable tourism, that perfect practice of mobility and exchange, to emerge as one of the world's largest single industries.

The new technologies affect the social relations of both sexuality and of reproduction, and not always in the same ways. The close ties of sexuality and instrumentality, of views of the body as a kind of private satisfaction- and utility-maximizing machine, are described nicely in sociobiological origin stories that stress a genetic calculus and explain the inevitable dialectic of domination of male and female gender roles.[23] These sociobiological stories depend on a high-tech view of the body as a biotic component or cybernetic communications system. Among the many transformations of reproductive situations is the medical one, where women's bodies have boundaries newly permeable to both "visualization" and "intervention." Of course, who controls the interpretation of bodily boundaries in medical hermeneutics is a major feminist issue. The speculum served as an icon of women's claiming their bodies in the 1970s; that handcraft tool is inadequate to express our needed body politics in the negotiation of reality in the practices of cy-borg reproduction. Self-help is not enough. The technologies of visualization recall the important cultural practice of hunting with the camera and the deeply predatory nature of a photo-graphic consciousness.[24] Sex, sexuality, and reproduction are central actors in high-tech myth systems structuring our imag-inations of personal and social possibility.

Another critical aspect of the social relations of the new technologies is the reformulation of expectations, culture, work, and reproduction for the large scientific and technical workforce. A major social and political danger is the forma-tion of a strongly bimodal social structure, with the masses of women and men of all ethnic groups, but especially people of color, confined to a homework economy, illiteracy of several varieties, and general redundancy and impotence, controlled by high-tech repressive apparatuses ranging from entertain-ment to surveillance and disappearance. An adequate socialist-feminist politics should address women in the privileged oc-cupational categories, and particularly in the production of science and technology that constructs scientific-technical discourses, processes, and objects.[25]

This issue is only one aspect of inquiry into the possibility of a feminist science, but it is important. What kind of consti-tutive role in the production of knowledge, imagination, and practice can new groups doing science have? How can these groups be allied with progressive social and political move-ments? What kind of political accountability can be con-

*A Cyborg Manifesto*

structed to tie women together across the scientific-technical hierarchies separating us? Might there be ways of developing feminist science/technology politics in alliance with antimilitary science facility conversion action groups? Many scientific and technical workers in Silicon Valley, the high-tech cowboys included, do not want to work on military science.[26] Can these personal preferences and cultural tendencies be welded into progressive politics among this professional middle class in which women, including women of color, are coming to be fairly numerous?

## WOMEN IN THE INTEGRATED CIRCUIT

Let me summarize the picture of women's historical locations in advanced industrial societies, as these positions have been restructured partly through the social relations of science and technology. If it was ever possible ideologically to characterize women's lives by the distinction of public and private domains—suggested by images of the division of working-class life into factory and home, of bourgeois life into market and home, and of gender existence into personal and political realms—it is now a totally misleading ideology, even to show how both terms of these dichotomies construct each other in practice and in theory. I prefer a network ideological image, suggesting the profusion of spaces and identities and the permeability of boundaries in the personal body and in the body

*A Cyborg Manifesto*

politic. "Networking" is both a feminist practice and a multinational corporate strategy—weaving is for oppositional cyborgs.

So let me return to the earlier image of the informatics of domination and trace one vision of women's "place" in the integrated circuit, touching only a few idealized social locations seen primarily from the point of view of advanced capitalist societies: Home, Market, Paid Workplace, State, School, Clinic-Hospital, and Church. Each of these idealized spaces is logically and practically implied in every other locus, perhaps analogous to a holographic photograph. I want to suggest the impact of the social relations mediated and enforced by the new technologies in order to help formulate needed analysis and practical work. However, there is no "place" for women in these networks, only geometries of difference and contradiction crucial to women's cyborg identities. If we learn how to read these webs of power and social life, we might learn new couplings, new coalitions. There is no way to read the following list from a standpoint of "identification," of a unitary self. The issue is dispersion. The task is to survive in the diaspora.

> *Home:* Women-headed households, serial monogamy, flight of men, old women alone, technology of domestic work, paid homework, reemergence of home sweatshops, home-based businesses and telecommuting, electronic cottage industry, urban homelessness, migration, module architecture, reinforced (simulated) nuclear family, intense domestic violence.

### A Cyborg Manifesto

*Market:* Women's continuing consumption work, newly targeted to buy the profusion of new production from the new technologies (especially as the competitive race among industrialized and industrializing nations to avoid dangerous mass unemployment necessitates finding ever bigger new markets for ever less clearly needed commodities); bimodal buying power, coupled with advertising targeting of the numerous affluent groups and neglect of the previous mass markets; growing importance of informal markets in labor and commodities parallel to high-tech, affluent market structures; surveillance systems through electronic funds transfer; intensified market abstraction (commodification) of experience, resulting in ineffective utopian or equivalent cynical theories of community; extreme mobility (abstraction) of marketing/financing systems; interpenetration of sexual and labor markets; intensified sexualization of abstracted and alienated consumption.

*Paid Workplace:* Continued intense sexual and racial division of labor, but considerable growth of membership in privileged occupational categories for many white women and people of color; impact of new technologies on women's work in clerical, service, manufacturing (especially textiles), agriculture, electronics; international restructuring of the working classes; development of new time arrangements to facilitate the homework economy (flex time, part time, over time, no time); homework and out work; increased pressures for two-tiered wage structures; significant numbers of

## A Cyborg Manifesto

people in cash-dependent populations worldwide with no experience or no further hope of stable employment; most labor "marginal" or "feminized."

*State:* Continued erosion of the welfare state; decentralizations with increased surveillance and control; citizenship by telematics; imperialism and political power broadly in the form of information-rich/information-poor differentiation; increased high-tech militarization increasingly opposed by many social groups; reduction of civil service jobs as a result of the growing capital intensification of office work, with implications for occupational mobility for women of color; growing privatization of material and ideological life and culture; close integration of privatization and militarization, the high-tech forms of bourgeois capitalist personal and public life; invisibility of different social groups to each other, linked to psychological mechanisms of belief in abstract enemies.

*School:* Deepening coupling of high-tech capital needs and public education at all levels, differentiated by race, class, and gender; managerial classes involved in educational reform and funding at the cost of remaining progressive educational democratic structures for children and teachers; education for mass ignorance and repression in technocratic and militarized culture; growing anti-science mystery cults in dissenting and radical political movements; continued relative scientific illiteracy among white women and people of color; growing industrial direction of education (especially higher education) by science-based multinationals (particularly in electronics- and biotechnology-dependent companies); highly educated, numerous elites in a progressively bimodal society.

*Clinic-Hospital:* Intensified machine–body relations; renegotiations of public metaphors that channel personal experience of the body, particularly in relation to reproduction, immune system functions, and "stress" phenomena; intensification of reproductive politics in response to world historical implications of women's unrealized, potential control of their relation to reproduction; emergence of new, historically specific diseases; struggles over meanings and means of health in environments pervaded by high-technology products and processes; continuing feminization of health work; intensified struggle over state responsibility for health; continued ideological role of popular health movements as a major form of American politics.

*Church:* Electronic fundamentalist "super-saver" preachers solemnizing the union of electronic capital and automated fetish gods; intensified importance of churches in resisting the militarized state; central struggle over women's meanings and authority in religion; continued relevance of spirituality, intertwined with sex and health, in political struggle.

The only way to characterize the informatics of domination is as a massive intensification of insecurity and cultural impoverishment, with common failure of subsistence networks for the most

*A Cyborg Manifesto*

vulnerable. Since much of this picture interweaves with the social relations of science and technology, the urgency of a socialist-feminist politics addressed to science and technology is plain. There is much now being done, and the grounds for political work are rich. For example, the efforts to develop forms of collective struggle for women in paid work, like SEIU's District 925,[27] should be a high priority for all of us. These efforts are profoundly tied to technical restructuring of labor processes and reformations of working classes. These efforts also are providing understanding of a more comprehensive kind of labor organization, involving community, sexuality, and family issues never privileged in the largely white male industrial unions.

The structural rearrangements related to the social relations of science and technology evoke strong ambivalence. But it is not necessary to be ultimately depressed by the implications of late-twentieth-century women's relation to all aspects of work, culture, production of knowledge, sexuality, and reproduction. For excellent reasons, most Marxisms see domination best and have trouble understanding what can only look like false consciousness and people's complicity in their own domination in late capitalism. It is crucial to remember that what is lost, perhaps especially from women's points of view, is often virulent forms of oppression, nostalgically naturalized in the face of current violation. Ambivalence toward the disrupted unities mediated by high-tech culture requires not sorting consciousness into categories of "clear-sighted critique grounding a solid political epistemology" versus "manipulated false consciousness," but subtle understanding of emerging pleasures, experiences, and powers with serious potential for changing the rules of the game.

There are grounds for hope in the emerging bases for new kinds of unity across race, gender, and class, as these elementary units of socialist-feminist analysis themselves suffer protean transformations. Intensifications of hardship experienced worldwide in connection with the social relations of science and technology are severe. But what people are experiencing is not transparently clear, and we lack sufficiently subtle connections for collectively building effective theories of experience. Present efforts—Marxist, psychoanalytic, feminist, anthropological—to clarify even "our" experience are rudimentary.

I am conscious of the odd perspective provided by my historical position—a PhD in biology for an Irish Catholic girl was made possible by Sputnik's impact on U.S. national science-education policy. I have a body and mind as much constructed by the post–Second World War arms race and Cold War as by the women's movements. There are more grounds for hope in focusing on the contradictory effects of politics designed to produce loyal American technocrats, which also produced large numbers of dissidents, than in focusing on the present defeats.

The permanent partiality of feminist points of view has consequences for our expectations of forms of political organiza-

*A Cyborg Manifesto*

tion and participation. We do not need a totality in order to work well. The feminist dream of a common language, like all dreams for a perfectly true language, of perfectly faithful naming of experience, is a totalizing and imperialist one. In that sense, dialectics too is a dream language, longing to resolve contradiction. Perhaps, ironically, we can learn from our fusions with animals and machines how not to be Man, the embodiment of Western logos. From the point of view of pleasure in these potent and taboo fusions, made inevitable by the social relations of science and technology, there might indeed be a feminist science.

## CYBORGS: A MYTH OF POLITICAL IDENTITY

I want to conclude with a myth about identity and boundaries that might inform late-twentieth-century political imaginations. I am indebted in this story to writers like Joanna Russ, Samuel R. Delany, John Varley, James Tiptree Jr., Octavia Butler, Monique Wittig, and Vonda McIntyre.[28] These are our storytellers exploring what it means to be embodied in high-tech worlds. They are theorists for cyborgs. Exploring conceptions of bodily boundaries and social order, the anthropologist Mary Douglas (1966, 1970) should be credited with helping us to con-

*A Cyborg Manifesto*

sciousness about how fundamental body imagery is to world-view, and so to political language.

French feminists like Luce Irigaray and Monique Wittig, for all their differences, know how to write the body; how to weave eroticism, cosmology, and politics from imagery of embodiment, and especially for Wittig, from imagery of fragmentation and reconstitution of bodies.[29] American radical feminists like Susan Griffin, Audre Lorde, and Adrienne Rich have profoundly affected our political imaginations — and perhaps restricted too much what we allow as a friendly body and political language.[30] They insist on the organic, opposing it to the technological. But their symbolic systems and the related positions of ecofeminism and feminist paganism, replete with organicisms, can only be understood in Sandoval's terms as oppositional ideologies fitting the late twentieth century. They would simply bewilder anyone not preoccupied with the machines and consciousness of late capitalism. In that sense they are part of the cyborg world. But there are also great riches for feminists in explicitly embracing the possibilities inherent in the breakdown of clean distinctions between organism and machine and similar distinctions structuring the Western self. It is the simultaneity of breakdowns that cracks the matrices of domination and opens geometric possibilities. What might be learned from personal and political "technological" pollution? I look briefly at two overlapping groups of texts for their insight into the construction of a potentially helpful cyborg myth: constructions of women of color and monstrous selves in feminist science fiction.

Earlier I suggested that "women of color" might be understood as a cyborg identity, a potent subjectivity synthesized from fusions of "outsider" identities, sedimented in the complex political-historical layerings of Audre Lorde's "biomythography," *Zami* (Lorde 1982; King 1987a, 1987b). There are material and cultural grids mapping this potential. Lorde (1984) captures the tone in the title of her *Sister Outsider*. In my political myth, Sister Outsider is the offshore woman, whom U.S. workers, female and feminized, are supposed to regard as the enemy preventing their solidarity, threatening their security. Onshore, inside the boundary of the United States, Sister Outsider is a potential amid the races and ethnic identities of women manipulated for division, competition, and exploitation in the same industries. "Women of color" are the preferred labor force for the science-based industries, the real women for whom the worldwide sexual market, labor market, and politics of reproduction kaleidoscope into daily life. Young Korean women hired in the sex industry and in electronics assembly are recruited from high schools, educated for the integrated circuit. Literacy, especially in English, distinguishes the "cheap" female labor so attractive to the multinationals.

Contrary to orientalist stereotypes of the "oral primitive," literacy is a special mark of women of color, acquired by U.S.

black women as well as men through a history of risking death to learn and to teach reading and writing. Writing has a special significance for all colonized groups. Writing has been crucial to the Western myth of the distinction between oral and written cultures, primitive and civilized mentalities, and more recently to the erosion of that distinction in "postmodernist" theories attacking the phallogocentrism of the West, with its worship of the monotheistic, phallic, authoritative, and singular work, the unique and perfect name.[31] Contests for the meanings of writing are a major form of contemporary political struggle. Releasing the play of writing is deadly serious. The poetry and stories of U.S. women of color are repeatedly about writing, about access to the power to signify; but this time that power must be neither phallic nor innocent. Cyborg writing must not be about the Fall, the imagination of a once-upon-a-time wholeness before language, before writing, before Man. Cyborg writing is about the power to survive, not on the basis of original innocence, but on the basis of seizing the tools to mark the world that marked them as other.

The tools are often stories, retold stories, versions that reverse and displace the hierarchical dualisms of naturalized identities. In retelling origin stories, cyborg authors subvert the central myths of origin of Western culture. We have all been colonized by those origin myths, with their longing for fulfillment in apocalypse. The phallogocentric origin stories most crucial for feminist cyborgs are built into the literal technolo-

*A Cyborg Manifesto*

gies—technologies that write the world, biotechnology and microelectronics—that have recently textualized our bodies as code problems on the grid of $C^3I$. Feminist cyborg stories have the task of recoding communication and intelligence to subvert command and control.

Figuratively and literally, language politics pervade the struggles of women of color; and stories about language have a special power in the rich contemporary writing by U.S. women of color. For example, retellings of the story of the indigenous woman Malinche, mother of the mestizo "bastard" race of the new world, master of languages, and mistress of Cortes, carry special meaning for Chicana constructions of identity. Cherríe Moraga in *Loving in the War Years* (1983) explores the themes of identity when one never possessed the original language, never told the original story, never resided in the harmony of legitimate heterosexuality in the garden of culture, and so cannot base identity on a myth or a fall from innocence and right to natural names, mother's or father's.[32] Moraga's writing, her superb literacy, is presented in her poetry as the same kind of violation as Malinche's mastery of the conqueror's language—a violation, an illegitimate production, that allows survival. Moraga's language is not "whole"; it is self-consciously spliced, a chimera of English and Spanish, both conquerors' languages. But it is this chimeric monster, without claim to an original language before violation, that crafts the erotic, competent, potent identities of women of color. *Sister Outsider* hints at the possibility of world survival not because of her innocence but because of her ability to live on the boundaries, to write without the founding myth of original wholeness, with its inescapable apocalypse of final return to a deathly oneness that Man has imagined to be the innocent and all-powerful Mother, freed at the End from another spiral of appropriation by her son. Writing marks Moraga's body, affirms it as the body of a woman of color, against the possibility of passing into the unmarked category of the Anglo father or into the orientalist myth of "original illiteracy" of a mother that never was. Malinche was mother here, not Eve before eating the forbidden fruit. Writing affirms Sister Outsider, not the Woman-before-the-Fall-into-Writing needed by the phallogocentric Family of Man.

Writing is preeminently the technology of cyborgs, etched surfaces of the late twentieth century. Cyborg politics are the struggle for language and the struggle against perfect communication, against the one code that translates all meaning perfectly, the central dogma of phallogocentrism. That is why cyborg politics insist on noise and advocate pollution, rejoicing in the illegitimate fusions of animal and machine. These are the couplings that make Man and Woman so problematic, subverting the structure of desire, the force imagined to generate language and gender, and so subverting the structure and modes of reproduction of "Western" identity, of nature and culture, of mirror and eye, slave and master, body and mind. "We" did not originally choose to be cyborgs, but choice grounds a liberal

*A Cyborg Manifesto*

politics and epistemology that imagine the reproduction of individuals before the wider replications of "texts."

From the perspective of cyborgs, freed of the need to ground politics in "our" privileged position of the oppression that incorporates all other dominations, the innocence of the merely violated, the ground of those closer to nature, we can see powerful possibilities. Feminisms and Marxisms have run aground on Western epistemological imperatives to construct a revolutionary subject from the perspective of a hierarchy of oppressions and/or a latent position of moral superiority, innocence, and greater closeness to nature. With no available original dream of a common language or original symbiosis promising protection from hostile "masculine" separation, but written into the play of a text that has no finally privileged reading or salvation history, to recognize "oneself" as fully implicated in the world, frees us of the need to root politics in identification, vanguard parties, purity, and mothering. Stripped of identity, the "bastard" race teaches about the power of the margins and the importance of a mother like Malinche. Women of color have transformed her from the evil mother of masculinist fear into the originally literate mother who teaches survival.

This is not just literary deconstruction, but liminal transformation. Every story that begins with original innocence and privileges the return to wholeness imagines the drama of life to be individuation, separation, the birth of the self, the tragedy of autonomy, the fall into writing, alienation—that is, war, tempered by imaginary respite in the bosom of the Other. These plots are ruled by a reproductive politics — rebirth without flaw, perfection, abstraction. In this plot women are imagined either better or worse off, but all agree they have less selfhood, weaker individuation, more fusion to the oral, to Mother, less at stake in masculine autonomy. But there is another route to having less at stake in masculine autonomy, a route that does not pass through Woman, Primitive, Zero, the Mirror Stage and its imaginary. It passes through women and other present-tense, illegitimate cyborgs, not of Woman born, who refuse the ideological resources of victimization so as to have a real life. These cyborgs are the people who refuse to disappear on cue, no matter how many times a "Western" commentator remarks on the sad passing of another primitive, another organic group done in by "Western" technology, by writing.[33] These real-life cyborgs (for example, the Southeast Asian village women workers in Japanese and U.S. electronics firms described by Aihwa Ong) are actively rewriting the texts of their bodies and societies.[34] Survival is at stake in this play of readings.

To recapitulate, certain dualisms have been persistent in Western traditions; they have all been systemic to the logics and practices of domination of women, people of color, nature, workers, animals — in short, domination of all constituted as others, whose task is to mirror the self. Chief among these troubling dualisms are self/other, mind/body, culture/nature, male/female, civilized/primitive, reality/appearance,

*A Cyborg Manifesto*

whole/part, agent/resource, maker/made, active/passive, right/wrong, truth/illusion, total/partial, God/man. The self is the One who is not dominated, who knows that by the service of the other, the other is the one who holds the future, who knows that by the experience of domination, which gives the lie to the autonomy of the self. To be One is to be autonomous, to be powerful, to be God; but to be One is to be an illusion, and so to be involved in a dialectic of apocalypse with the other. Yet to be other is to be multiple, without clear boundary, frayed, insubstantial. One is too few, but two are too many.

High-tech culture challenges these dualisms in intriguing ways. It is not clear who makes and who is made in the relation between human and machine. It is not clear what is mind and what is body in machines that resolve into coding practices. Insofar as we know ourselves in both formal discourse (for example, biology) and in daily practice (for example, the homework economy in the integrated circuit), we find ourselves to be cyborgs, hybrids, mosaics, chimeras. Biological organisms have become biotic systems, communications devices like others. There is no fundamental, ontological separation in our formal knowledge of machine and organism, of technical and organic. The replicant Rachel in the Ridley Scott film *Blade Runner* stands as the image of a cyborg culture's fear, love, and confusion.

One consequence is that our sense of connection to our tools is heightened. The trance state experienced by many computer users has become a staple of science-fiction film and cultural jokes. Perhaps paraplegics and other severely handicapped people can (and sometimes do) have the most intense experiences of complex hybridization with other communications devices.[35] Anne McCaffrey's prefeminist *The Ship Who Sang* (1969) explored the consciousness of a cyborg, hybrid of girl's brain and complex machinery, formed after the birth of a severely handicapped child. Gender, sexuality, embodiment, skill: all were reconstituted in the story. Why should our bodies end at the skin, or include at best other beings encapsulated by skin? From the seventeenth century till now, machines could be animated—given ghostly souls to make them speak or move or to account for their orderly development and mental capacities. Or organisms could be mechanized—reduced to body understood as resource of mind. These machine/organism relationships are obsolete, unnecessary. For us, in imagination and in other practice, machines can be prosthetic devices, intimate components, friendly selves. We don't need organic holism to give impermeable wholeness, the total woman and her feminist variants (mutants?). Let me conclude this point by a very partial reading of the logic of the cyborg monsters of my second group of texts, feminist science fiction.

The cyborgs populating feminist science fiction make very problematic the statuses of man or woman, human, artifact, member of a race, individual entity, or body. Katie King clarifies how pleasure in reading these fictions is not largely based on identification. Students facing Joanna Russ for the first time,

*A Cyborg Manifesto*

students who have learned to take modernist writers like James Joyce or Virginia Woolf without flinching, do not know what to make of *The Adventures of Alyx* or *The Female Man,* where characters refuse the reader's search for innocent wholeness while granting the wish for heroic quests, exuberant eroticism, and serious politics. *The Female Man* is the story of four versions of one genotype, all of whom meet, but even taken together do not make a whole, resolve the dilemmas of violent moral action, or remove the growing scandal of gender. The feminist science fiction of Samuel R. Delany, especially *Tales of Nevèrÿon,* mocks stories of origin by redoing the neolithic revolution, replaying the founding moves of Western civilization to subvert their plausibility. James Tiptree Jr., an author whose fiction was regarded as particularly manly until her "true" gender was revealed, tells tales of reproduction based on nonmammalian technologies like alternation of generations of male brood pouches and male nurturing. John Varley constructs a supreme cyborg in his arch-feminist exploration of Gaea, a mad goddess-planet-trickster-old woman-technological-device on whose surface an extraordinary array of post-cyborg symbioses are spawned. Octavia Butler writes of an African sorceress pitting her powers of transformation against the genetic manipulations of her rival (*Wild Seed*), of time warps that bring a modern U.S. black woman into slavery where her actions in relation to her white master–ancestor determine the possibility of her own birth (*Kindred*), and of the illegitimate insights into identity and community of an adopted cross-species child who came to know the enemy as self (*Survivor*). In *Dawn* (1987), the first installment of a series called *Xenogenesis*, Butler tells the story of Lilith Iyapo, whose personal name recalls Adam's first and repudiated wife and whose family name marks her status as the widow of the son of Nigerian immigrants to the United States. A black woman and a mother whose child is dead, Lilith mediates the transformation of humanity through genetic exchange with extraterrestrial lovers/rescuers/destroyers/genetic engineers, who re-form Earth's habitats after the nuclear holocaust and coerce surviving humans into intimate fusion with them. It is a novel that interrogates reproductive, linguistic, and nuclear politics in a mythic field structured by late-twentieth-century race and gender.

Because it is particularly rich in boundary transgressions, Vonda McIntyre's *Superluminal* can close this truncated catalogue of promising and dangerous monsters who help redefine the pleasures and politics of embodiment and feminist writing. In a fiction where no character is "simply" human, human status is highly problematic. Orca, a genetically altered diver, can speak with killer whales and survive deep ocean conditions, but she longs to explore space as a pilot, necessitating bionic implants jeopardizing her kinship with the divers and cetaceans. Transformations are effected by virus vectors carrying a new developmental code, by transplant surgery, by implants of microelectronic devices, by analogue doubles, and other means.

## A Cyborg Manifesto

Laenea becomes a pilot by accepting a heart implant and a host of other alterations allowing survival in transit at speeds exceeding that of light. Radu Dracul survives a virus-caused plague in his outerworld planet to find himself with a time sense that changes the boundaries of spatial perception for the whole species. All the characters explore the limits of language; the dream of communicating experience; and the necessity of limitation, partiality, and intimacy even in this world of protean transformation and connection. *Superluminal* stands also for the defining contradictions of a cyborg world in another sense; it embodies textually the intersection of feminist theory and colonial discourse in the science fiction I have alluded to in this essay. This is a conjunction with a long history that many "First World" feminists have tried to repress, including myself in my readings of *Superluminal* before being called to account by Zoë Sofoulis (n.d.), whose different location in the world system's informatics of domination made her acutely alert to the imperialist moment of all science fiction cultures, including women's science fiction. From an Australian feminist sensitivity, Sofoulis remembered more readily McIntyre's role as writer of the adventures of Captain Kirk and Spock in TV's *Star Trek* series than her rewriting the romance in *Superluminal*.

Monsters have always defined the limits of community in Western imaginations. The Centaurs and Amazons of ancient Greece established the limits of the centered polis of the Greek male human by their disruption of marriage and boundary pollutions of the warrior with animality and woman. Unseparated twins and hermaphrodites were the confused human material in early modern France who grounded discourse on the natural and supernatural, medical and legal, portents and diseases — all crucial to establishing modern identity.[36] In the evolutionary and behavioral sciences, monkeys and apes have marked the multiple boundaries of late-twentieth-century industrial identities. Cyborg monsters in feminist science fiction define quite different political possibilities and limits from those proposed by the mundane fiction of Man and Woman.

There are several consequences to taking seriously the imagery of cyborgs as other than our enemies. Our bodies, ourselves; bodies are maps of power and identity. Cyborgs are no exception. A cyborg body is not innocent; it was not born in a garden; it does not seek unitary identity and so generate antagonistic dualisms without end (or until the world ends); it takes irony for granted. One is too few, and two is only one possibility. Intense pleasure in skill, machine skill, ceases to be a sin, but an aspect of embodiment. The machine is not an *it* to be animated, worshipped, and dominated. The machine is us, our processes, an aspect of our embodiment. We can be responsible for machines; *they* do not dominate or threaten us. We are responsible for boundaries; we are they. Up till now (once upon a time), female embodiment seemed to be given, organic, necessary; and female embodiment seemed to mean skill in mothering and its metaphoric extensions. Only by being out of place could we

*A Cyborg Manifesto*

take intense pleasure in machines, and then with excuses that this was organic activity after all, appropriate to females. Cyborgs might consider more seriously the partial, fluid, sometimes aspect of sex and sexual embodiment. Gender might not be global identity after all, even if it has profound historical breadth and depth.

The ideologically charged question of what counts as daily activity, as experience, can be approached by exploiting the cyborg image. Feminists have recently claimed that women are given to dailiness, that women more than men somehow sustain daily life and so have a privileged epistemological position potentially. There is a compelling aspect to this claim, one that makes visible unvalued female activity and names it as the ground of life.

But *the* ground of life? What about all the ignorance of women, all the exclusions and failures of knowledge and skill? What about men's access to daily competence, to knowing how to build things, to take them apart, to play? What about other embodiments? Cyborg gender is a local possibility taking a global vengeance. Race, gender, and capital require a cyborg theory of wholes and parts. There is no drive in cyborgs to produce total theory, but there is an intimate experience of boundaries, their construction and deconstruction. There is a myth system waiting to become a political language to ground one way of looking at science and technology and challenging the informatics of domination—in order to act potently.

*A Cyborg Manifesto*

66

One last image: organisms and organismic, holistic politics depend on metaphors of rebirth and invariably call on the resources of reproductive sex. I would suggest that cyborgs have more to do with regeneration and are suspicious of the reproductive matrix and of most birthing. For salamanders, regeneration after injury, such as the loss of a limb, involves regrowth of structure and restoration of function with the constant possibility of twinning or other odd topographical productions at the site of former injury. The regrown limb can be monstrous, duplicated, potent. We have all been injured, profoundly. We require regeneration, not rebirth, and the possibilities for our reconstitution include the utopian dream of the hope for a monstrous world without gender.

Cyborg imagery can help express two crucial arguments in this essay: first, the production of universal, totalizing theory is a major mistake that misses most of reality, probably always, but certainly now; and second, taking responsibility for the social relations of science and technology means refusing an antiscience metaphysics, a demonology of technology, and so means embracing the skillful task of reconstructing the boundaries of daily life, in partial connection with others, in communication with all of our parts. It is not just that science and technology are possible means of great human satisfaction, as well as a matrix of complex dominations. Cyborg imagery can suggest a way out of the maze of dualisms in which we have explained our bodies and our tools to ourselves. This is a dream not of a common language, but of a powerful infidel heteroglossia. It is an imagination of a feminist speaking in tongues to strike fear into the circuits of the supersavers of the new right. It means both building and destroying machines, identities, categories, relationships, space stories. Though both are bound in the spiral dance, I would rather be a cyborg than a goddess.

## NOTES

1. See Zoe Sofoulis (n.d.).

2. See Hilary Klein 1989.

3. Useful references to left and/or feminist radical science movements and theory and to biological/biotechnical issues include Bleier 1984, 1986; Harding 1986; Fausto-Sterling 1985; Gould 1981; Hubbard et al. 1979; Keller 1985; Lewontin et al. 1984. See also *Radical Science Journal* (which became *Science as Culture* in 1987): 26 Freegrove Road, London N7 9RQ; and *Science for the People,* 897 Main Street, Cambridge, Massachusetts 02139.

4. Starting points for left and/or feminist approaches to technology and politics include Cowan 1983, 1986; Rothschild 1983; Traweek 1988; Young and Levidow 1981, 1985; Weisenbaum 1976; Winner 1977, 1986; Zimmerman 1983; Athanasiou 1987; Cohn 1987a, 1987b; Winograd and Flores 1986; Edwards 1985. *Global Electronics Newsletter,* 867 West Dana Street, #204, Mountain View, California 94041; *Processed World,* 55 Sutter Street, San Francisco, California 94104; ISIS, Women's International Information and Communication Service, P.O. Box 50 (Cornavin), 1211 Geneva 2, Switzerland; and Via Santa Maria Dell'Anima 30, 00186 Rome, Italy. Fundamental approaches to modern social studies of science that do

*A Cyborg Manifesto*

not continue the liberal mystification that all started with Thomas Kuhn include Knorr-Cetina 1981; Knorr-Cetina and Mulkay 1983; Latour and Woolgar 1979; Young 1979. The 1984 Directory of the Network for the Ethnographic Study of Science, Technology, and Organization lists a wide range of people and projects crucial to better radical analysis, available from NESSTO, P.O. Box 11442, Stanford, California 94305.

5. A provocative, comprehensive argument about the politics and theories of "postmodernism" is made by Fredric Jameson (1984), who argues that postmodernism is not an option, a style among others, but a cultural dominant requiring radical reinvention of left politics from within; there is no longer any place from without that gives meaning to the comforting fiction of critical distance. Jameson also makes clear why one cannot be for or against postmodernism, an essentially moralist move. My position is that feminists (and others) need continuous cultural reinvention, most modernist critique, and historical materialism; only a cyborg would have a chance. The old dominations of white capitalist patriarchy seem nostalgically innocent now: they normalized heterogeneity, into man and woman, white and black, for example. "Advanced Capitalism" and postmodernism release heterogeneity without a norm, and we are flattened, without subjectivity, which requires depth, even unfriendly and drowning depths. It is time to write *The Death of the Clinic*. The clinic's methods required bodies and works; we have texts and surfaces. Our dominations don't work by medicalization and normalization anymore; they work by networking, communications redesign, stress management. Normalization gives way to automation, utter redundancy. Michel Foucault's *Birth of the Clinic* (1963), *History of Sexuality* (1976), and *Discipline and Punish* (1975) name a form of power at its moment of implosion. The discourse of biopolitics gives way to technobabble, the language of the spliced substantive; no noun is left whole by the multinationals. These are their

## A Cyborg Manifesto

names, listed from one issue of *Science*: Tech-Knowledge, Genentech, Allergen, Hybritech, Compupro, Genen-cor, Syntex, Allelix, Agrigenetics Corp., Syntro, Codon, Repligen, Micro/Angelo from Scion Corp., Percom Data, Inter Systems, Cyborg Corp., Statcom Corp., Intertec. If we are imprisoned by language, then escape from that prison-house requires language poets, a kind of cultural restriction enzyme to cut the code; cyborg heteroglossia is one form of radical cultural politics. For cyborg poetry see Perloff 1984; Fraser 1984. For feminist modernist/postmodernist cyborg writing, see *HOW(ever)*, 971 Corbett Avenue, San Francisco, California 94131.

6. The U.S. equivalent of Mills and Boon.

7. Baudrillard 1983 and Jameson 1984 (page 66) point out that Plato's definition of the simulacrum is the copy for which there is no original, i.e., the world of advanced capitalism, of pure exchange. See *Discourse* 9 (Spring/Summer 1987) for a special issue on technology (cybernetics, ecology, and the postmodern imagination).

8. A practice at once both spiritual and political that linked guards and arrested antinuclear demonstrators in the Alameda County Jail in California in the early 1980s.

9. For ethnographic accounts and political evaluations, see Epstein 1993; Sturgeon 1986. Without explicit irony, adopting the spaceship earth/whole earth logo of the planet photographed from space, set off by the slogan "Love Your Mother," the May 1987 Mothers and Others Day action at the nuclear weapons testing facility in Nevada nonetheless took account of the tragic contradictions of views of the earth. Demonstrators applied for official permits to be on the land from officers of the Western Shoshone tribe, whose territory was invaded by the U.S. government when it built the nuclear weapons test ground in the 1950s. Arrested for trespassing, the demonstrators argued that the police and weapons facil-

ity personnel, without authorization from the proper officials, were the trespassers. One affinity group at the women's action called themselves the Surrogate Others; and in solidarity with the creatures forced to tunnel in the same ground with the bomb, they enacted a cyborgian emergence from the constructed body of a large, nonheterosexual desert worm. I was a member of that affinity group.

10. Powerful developments of coalition politics emerge from "Third World" speakers, speaking from nowhere, the displaced center of the universe, earth: "We live on the third planet from the sun" — *Sun Poem* by Jamaican writer Edward Kamau Braithwaite, review by Mackey 1984. Contributors to Smith 1983 ironically subvert naturalized identities precisely while constructing a place from which to speak called home. See especially Reagon (in Smith 1983, 356–68); Trinh T. Minh-ha 1986–87a, b.

11. See hooks 1981, 1984; Hull et al. 1982. Toni Cade Bambara (1981) wrote an extraordinary novel in which the women of color theater group the Seven Sisters explores a form of unity. See analysis by Butler-Evans 1987.

12. On orientalism in feminist works and elsewhere, see Lowe 1986; Said 1978; Mohanty 1984; *Many Voices, One Chant: Black Feminist Perspectives* (1984).

13. Katie King (1986, 1987a) has developed a theoretically sensitive treatment of the workings of feminist taxonomies as genealogies of power in feminist idealogy and polemic. King examines Jaggar's (1983) problematic example of taxonomizing feminisms to make a little machine producing the desired final position. My caricature here of socialist and radical feminism is also an example.

14. The central role of object relations versions of psychoanalysis and related strong universalizing moves in discussing reproduction, caring work, and mothering in many approaches to epistemology underline their

*A Cyborg Manifesto*

authors' resistance to what I am calling postmodernism. For me, both the universalizing moves and these versions of psychoanalysis make analysis of "women's place in the integrated circuit" difficult and lead to systematic difficulties in accounting for or even seeing major aspects of the construction of gender and gendered social life. The feminist standpoint argument has been developed by Flax 1983; Harding 1986; Harding and Hintikka 1983; Hartsock 1983a, 1983b; O'Brien 1981; H. Rose 1983; Smith 1974, 1979. For rethinking theories of feminist materialism and feminist standpoints in response to criticism, see Harding 1986, 163–96; Hartsock 1987; and S. Rose 1986.

15. I make an argumentative category error in "modifying" MacKinnon's positions with the qualifier "radical," thereby generating my own reductive critique of extremely hetergeneous writing, which does explicitly use that label, by my taxonomically interested argument about writing, which does not use the modifier and which brooks no limits and thereby adds to the various dreams of a common, in the sense of univocal, language for feminism. My category error was occasioned by an assignment to write from a particular taxonomic position that itself has a heterogeneous history, socialist-feminism, for *Socialist Review*, published in *SR* as "The Cyborg Manifesto." A critique indebted to MacKinnon, but without the reductionism and with an elegant feminist account of Foucault's paradoxical conservatism on sexual violence (rape), is de Lauretis 1985 (see also 1986, 1–19). A theoretically elegant feminist social-historical examination of family violence, which insists on women's, men's, and children's complex agency without losing sight of the material structures of male domination, race, and class, is Gordon 1988.

16. See Kristeva 1984.

17. This chart was published in 1985 in the "Cyborg Manifesto." My previous efforts to understand biology as a cybernetic command-control

discourse and organisms as "natural-technical objects of knowledge" were Haraway 1979, 1983, 1984. A later version, with a shifted argument, appears in Haraway 1989.

18. For progressive analyses and action on the biotechnology debates, see *GeneWatch, a Bulletin of the Committee for Responsible Genetics,* 5 Doane St., 4th Floor, Boston, Massachusetts 02109; Genetic Screening Study Group (formerly the Sociobiology Study Group of Science for the People), Cambridge, Massachusetts; Wright 1982, 1986; Yoxen 1983.

19. Starting references for "women in the integrated circuit": D'Onofrio-Flores and Pfafflin 1982; Fernandez-Kelly 1983; Fuentes and Ehrenreich 1983; Grossman 1980; Nash and Fernandez-Kelly 1983; A. Ong 1987; Science Policy Research Unit 1982.

20. For the "homework economy outside the home" and related arguments, see Gordon 1983; Gordon and Kimball 1985; Stacey 1987; Reskin and Hartmann 1986; *Women and Poverty* 1984; S. Rose 1986; Collins 1982; Burr 1982; Gregory and Nussbaum 1982; Piven and Coward 1982; Microelectronics Group 1980; Stallard et al. 1983, which includes a useful organization and resource list.

21. The conjunction of the Green Revolution's social relations with biotechnologies like plant genetic engineering makes the pressures on land in the Third World increasingly intense. The U.S. Agency for International Development's estimates used at the 1984 World Food Day are that in Africa women produce about 90 percent of rural food supplies, about 60–80 percent in Asia, and provide 40 percent of agricultural labor in the Near East and Latin America (*New York Times* 1984). Blumberg charges that world organizations' agricultural politics, as well as those of multinationals and national governments in the Third World, generally ignore fundamental issues in the sexual division of labor. The present tragedy of famine in Africa might owe as much to male supremacy as to

*A Cyborg Manifesto*

capitalism, colonialism, and rain patterns. More accurately, capitalism and racism are usually structurally male dominant. See also Blumberg 1981; Hacker 1984; Hacker and Bovit 1981; Busch and Lacy 1983; Wilfred 1982; Sachs 1983; International Fund for Agricultural Development 1985; Bird 1984.

22. See also Enloe 1983a, 1983b.

23. For a feminist version of this logic, see Hrdy 1981. For an analysis of scientific women's storytelling practices, especially in relation to sociobiology in evolutionary debates around child abuse and infanticide, see Haraway 1989.

24. For the moment of transition of hunting with guns to hunting with cameras in the construction of popular meanings of nature for an American urban immigrant public, see Haraway 1984–85, 1989; Nash 1979; Sontag 1977; Preston 1984.

25. For guidance for thinking about the political/cultural/racial implications of the history of women doing science in the United States, see Haas and Perucci 1984; Hacker 1981; Keller 1983; National Science Foundation 1988; Rossiter 1982; Schiebinger 1987; Haraway 1989.

26. See Markoff and Siegel 1983. High Technology Professionals for Peace and Computer Professionals for Social Responsibility are promising organizations.

27. The Service Employees International Union's office workers' organization in the United States.

28. See King 1984. An abbreviated list of feminist science fiction underlying themes of this essay: Octavia Butler, *Wild Seed, Mind of My Mind, Kindred, Survivor*; Suzy McKee Charnas, *Motherlines*; Samuel R. Delany, the Nevèrÿon series; Anne McCaffery, *The Ship Who Sang, Dinosaur Planet*; Vonda McIntyre, *Superluminal, Dreamsnake*; Joanna Russ,

*Adventures of Alix, The Female Man*; James Tiptree Jr., *Star Songs of an Old Primate, Up the Walls of the World*; John Varley, *Titan, Wizard, Demon.*

29. French feminisms contribute to cyborg heteroglossia: Burke 1981; Irigaray 1977, 1979; Marks and de Courtivron 1980; *Signs: Journal of Women in Culture and Society* 1981 (Autumn); Wittig 1973; Duchen 1986. For English translation of some currents of Francophone feminism, see *Feminist Issues: A Journal of Feminist Social and Political Theory* (1980).

30. But all these poets are very complex, not least in their treatment of themes of lying and erotic, decentered collective and personal identities: Griffin 1978; Lorde 1984; Rich 1978.

31. See Derrida 1976 (especially part II); Lévi-Strauss 1973 (especially "The Writing Lesson"); Gates 1985; Kahn and Neumaier 1985; Ong 1982; Kramarae and Treichler 1985.

32. The sharp relation of women of color to writing as theme and politics can be approached through the program for "The Black Woman and the Diaspora: Hidden Connections and Extended Acknowledgments," An International Literary Conference, Michigan State University, October 1985; Evans 1984; Christian 1985; Carby 1987; Fisher 1980; *Frontiers* 1980, 1983; Kingston 1976; Lerner 1973; Giddings 1985; Moraga and Anzaldúa 1981; Morgan 1984. Anglophone European and Euro-American women have also crafted special relations to their writing as a potent sign: Gilbert and Gubar 1979; Russ 1983.

33. The convention of ideologically taming militarized high technology by publicizing its applications to speech and motion problems of the disabled/differently abled takes on a special irony in monotheistic, patriarchal, and frequently anti-Semitic culture when computer-generated speech allows a boy with no voice to chant the Haftorah at his bar mitzvah. See Sussman 1986. Making the always context-relative social definitions

## A Cyborg Manifesto

of "ableness" particularly clear, military high-tech has a way of making human beings disabled by definition, a perverse aspect of much automated battlefield and Star Wars research and development. See Wilford 1986.

34. See A. Ong 1987.

35. James Clifford (1985, 1988) argues persuasively for recognition of continuous reinvention, the stubborn nondisappearance of those "marked" by Western imperializing practices.

36. See DuBois 1982; Daston and Mark n.d.; Park and Daston 1981. The noun *monster* shares its root with the verb *to demonstrate.*

## BIBLIOGRAPHY

Athanasiou, Tom. 1987. "High-Tech Politics: The Case of Artifical Intelligence." *Socialist Review* 92: 7–35.

Bambara, Toni Cade. 1981. *The Salt Eaters.* New York: Vintage/Random House.

Baudrillard, Jean. 1983. *Simulations.* Trans. P. Foss, P. Patton, and P. Beitchman. New York: Semiotext[e].

Bird, Elizabeth. 1984. "Green Revolution Imperialism, I and II." Papers delivered to the History of Consciousness Board, University of California, Santa Cruz.

Bleier, Ruth. 1984. *Science and Gender: A Critique of Biology and Its Themes on Women.* New York: Pergamon.

Blumberg, Rae Lessor. 1981. *Stratification: Socioeconomic and Sexual Inequality.* Boston: Little, Brown.

——. 1983. "A General Theory of Sex Stratification and Its Application to Positions of Women in Today's World Economy." Paper delivered to the Sociology Board of the University of California, Santa Cruz.

*A Cyborg Manifesto*

Burke, Carolyn. 1981. "Irigaray through the Looking Glass." *Feminist Studies* 7 (2): 288–306.

Burr, Sara G. 1982. "Women and Work." In *The Women's Annual, 1981,* ed. Barbara K. Haber. Boston: G. K. Hall.

Busch, Lawrence, and William Lacy. 1983. *Science, Agriculture, and the Politics of Research.* Boulder, Colo.: Westview Press.

Butler-Evans, Elliott. 1987. "Race, Gender and Desire: Narrative Strategies and the Production of Ideology in the Fiction of Toni Cade Bambara, Toni Morrison and Alice Walker." PhD diss., University of California, Santa Cruz.

Butler, Octavia. 1979. *Survivor.* New York: Signet.

——. 1984. *Mind of My Mind.* New York: Grand Central Publishing.

——. 2001. *Wild Seed.* New York: Grand Central Publishing.

——. 2003. *Kindred.* Boston: Beacon Press.

Carby, Hazel. 1987. *Reconstructing Womanhood: The Emergence of the Afro-American Woman Novelist.* New York: Oxford University Press.

Charnas, Suzy McKee. 1955. *Motherlines.* New York: Berkeley.

Christian, Barbara. 1985. *Black Feminist Criticism: Perspectives on Black Women Writers.* New York: Pergamon Press.

Clifford, James. 1985. "On Ethnographic Allegory." In *The Poetics and Politics of Ethnography,* ed. James Clifford and George Marcus. Berkeley: University of California Press.

——. 1988. *The Predicament of Culture: Twentieth-century Ethnography, Literature, and Art.* Cambridge, Mass.: Harvard University Press.

Cohn, Carol. 1987a. "Nuclear Language and How We Learned to Pat the Bomb." *Bulletin of Atomic Scientists* 43 (5): 17–24.

——. 1987b. "Sex and Death in the Rational World of Defense Intellectuals." *Signs* 12 (4): 687–718.

# A Cyborg Manifesto

Collins, Patricia Hill. 1982. "Third World Women in America." In *The Women's Annual, 1981,* ed. Barbara K. Haber. Boston: G. K. Hall.

Cowan, Ruth Schwartz. 1983. *More Work for Mother: The Ironies of Household Technology from the Open Hearth to the Microwave.* New York: Basic Books.

———, ed. 1986. *Feminist Approaches to Science.* New York: Pergamon Press.

Daston, Lorraine, and Katherin Park. N.d. "Hermaphrodites in Renaissance France." Unpublished manuscript.

Delany, Samuel R. 1979. *Tales of Nevèrÿon.* New York: Bantam Books.

de Lauretis, Teresa. 1985. "The Violence of Rhetoric: Considerations on Representation and Gender." *Semiotica* 54: 11–31.

———. 1986. "Feminist Studies/Critical Studies: Issues, Terms, and Contexts." In *Feminist Studies/Critical Studies,* ed. T. de Lauretis, 1–19. Bloomington: Indiana University Press.

Derrida, Jacques. 1976. *Of Grammatology.* Trans. G. C. Spivak. Baltimore: Johns Hopkins University Press.

de Waal, Frans. 1982. *Chimpanzee Politics: Power and Sex among Apes.* New York: Harper and Row.

D'Onofrio-Flores, Pamela, and Sheila M. Pfafflin, eds. 1982. *Scientific-Technological Change and the Role of Women in Development.* Boulder, Colo.: Westview Press.

Douglas, Mary. 1966. *Purity and Danger.* London: Routledge and Kegan Paul.

———. 1970. *Natural Symbols.* London: Cresset Press.

DuBois, Page. 1982. *Centaurs and Amazons.* Ann Arbor: University of Michigan Press.

Duchen, Claire. 1986. *Feminism in France from May '68 to Mitterand.* London: Routledge and Kegan Paul.

## A Cyborg Manifesto

Edwards, Paul. 1985. "Border Wars: The Science and Politics of Artificial Intelligence." *Radical America* (19) 6: 39–52.

Enloe, Cynthia. 1983a. "Women Textile Workers in the Militarization of Southeast Asia." In Nash and Fernandez-Kelly 1983, 407–25.

——. 1983b. *Does Khaki Become You? The Militarisation of Women's Lives.* Boston: South End Press.

Epstein, Barbara. 1993. *Political Protest and Cultural Revolution: Nonviolent Direct Action in the Seventies and Eighties.* Berkeley: University of California Press.

Evans, Mari, ed. 1984. *Black Women Writers: A Critical Evaluation.* Garden City, N.Y.: Doubleday/Anchor.

Fausto-Sterling, Anne. 1985. *Myths of Gender: Biological Theories about Women and Men.* New York: Basic Books.

*Feminist Issues: A Journal of Feminist Social and Political Theory.* 1980. 1 (1): special issue on Francophone feminisms.

Fernandez-Kelly, Maria Patricia. 1983. *For We Are Sold, I and My People.* Albany: State University of New York Press.

Fisher, Dexter, ed. 1980. *The Third Woman: Minority Women Writers of the United States.* Boston: Houghton Mifflin.

Flax, Jane. 1983. "Political Philosophy and the Patriarchal Unconscious: A Psychoanalytic Perspective on Epistemology and Metaphysics." In Harding and Hintikka 1983, 245–82.

Foucault, Michel. 1963. *The Birth of the Clinic: An Archaeology of Medical Perception.* Trans. A. M. Smith. New York: Vintage.

——. 1975. *Discipline and Punish: The Birth of the Prison.* Trans. Alan Sheridan. New York, Vintage.

——. 1976. *The History of Sexuality, Vol. 1: An Introduction.* Trans. Robert Hurley. New York: Pantheon, 1978.

## A Cyborg Manifesto

Fraser, Kathleen. 1984. *Something. Even Human Voices. In the Foreground, a Lake.* Berkeley, Calif.: Kelsey St. Press.

*Frontiers: A Journal of Women's Studies.* 1980. Volume 1.

——. 1983. Volume 3.

Fuentes, Annette, and Barbara Ehrenreich. 1983. *Women in the Global Factory.* Boston: South End Press.

Gates, Henry Louis Jr. 1985. "Writing 'Race' and the Difference It Makes." In *"Race," Writing and Difference* (special issue), *Critical Inquiry* 12 (1): 1–20.

Giddings, Paula. 1985. *When and Where I Enter: The Impact of Black Women on Race and Sex in America.* Toronto: Bantam Books.

Gilbert, Sandra M., and Susan Gubar. 1979. *The Madwoman in the Attic: The Woman Writer and the Nineteenth-Century Literary Imagination.* New Haven, Conn.: Yale University Press.

Gordon, Linda. 1988. *Heroes of Their Own Lives: The Politics and History of Family Violence, Boston 1880–1960.* New York: Viking Penguin.

Gordon, Richard. 1983. "The Computerization of Daily Life, the Sexual Division of Labor, and the Homework Economy." Presented at the Silicon Valley Workshop Conference, University of California, Santa Cruz.

——, and Linda Kimball. 1985. "High-Technology, Employment and the Challenges of Education." Silicon Valley Research Project, Working Paper, no. 1.

Gould, Stephen Jay. 1981. *The Mismeasure of Man.* New York: W. W. Norton.

Gregory, Judith, and Karen Nussbaum. 1982. "Race against Time: Automation of the Office." *Office: Technology and People* 1: 197–236.

Griffin, Susan. 1978. *Women and Nature: The Roaring Inside Her.* New York: Harper and Row.

# A Cyborg Manifesto

Grossman, Rachel. 1980. "Women's Place in the Integrated Circuit." *Radical America* 14 (1): 29–50.

Haas, Violet, and Carolyn Perucci, eds. 1984. *Women in Scientific and Engineering Professions*. Ann Arbor: University of Michigan Press.

Hacker, Sally. 1981. "The Culture of Engineering: Women, Workplace, and Machine." *Women's Studies International Quarterly* 4 (3): 341–53.

———. 1984. "Doing It the Hard Way: Ethnographic Studies in the Agribusiness and Engineering Classroom." Presented at the California American Studies Association, Pomona.

———, and Liza Bovit. 1981. "Agriculture to Agribusiness: Technical Imperatives and Changing Roles." Presented at the Society for the History of Technology, Milwaukee.

Haraway, Donna J. 1979. "The Biological Enterprise: Sex, Mind, and Profit from Human Engineering to Sociobiology." *Radical History Review* 20: 206–37.

———. 1983. "Signs of Dominance: From a Physiology to a Cybernetics of Primate Society." *Studies in History of Biology* 6: 129–219.

———. 1984. "Class, Race, Sex, Scientific Objects of Knowledge: A Socialist-Feminist Perspective on the Social Construction of Productive Knowledge and Some Political Consequences." In Haas and Perucci 1984, 212–29.

———. 1984–85. "Teddy Bear Patriarchy: Taxidermy in the Garden of Eden, New York City, 1908–36." *Social Text* 11: 20–64.

———. 1989. *Primate Visions: Gender, Race, and Nature in the World of Modern Science*. New York: Routledge.

Harding, Sandra. 1978. "What Causes Gender Privilege and Class Privilege?" Presented at the American Philosophical Association.

———. 1983. "Why Has the Sex/Gender System Become Visible Only Now?" In Harding and Hintikka 1983, 311–24.

## A Cyborg Manifesto

———. 1986. *The Science Question in Feminism*. Ithaca, N.Y.: Cornell University Press.

———, and Merrill Hintikka, eds. 1983. *Discovering Reality: Feminist Perspectives on Epistemology, Metaphysics, Methodology, and Philosophy of Science*. Dordrecht, the Netherlands: D. Reidel.

Hartsock, Nancy. 1983a. "The Feminist Standpoint: Developing the Ground for a Specifically Feminist Historical Materialism." In Harding and Hintikka 1983, 283–310.

———. 1983b. *Money, Sex, and Power*. New York: Longman.

———. 1987. "Rethinking Modernism: Minority and Majority Theories." *Cultural Critique* 7: 187–206.

Hogness, Erik Rusten. 1983. "Why Stress? A Look at the Making of Stress, 1936-56." Unpublished manuscript.

hooks, bell. 1981. *Ain't I a Woman*. Boston: South End Press.

———. 1984. *Feminist Theory: From Margin to Center*. Boston: South End Press.

Hrdy, Sarah Blaffer. 1981. *The Woman That Never Evolved*. Cambridge, Mass.: Harvard University Press.

Hubbard, Ruth, and Marian Lowe, eds. 1979. *Genes and Gender*. Vol. 2, *Pitfalls in Research on Sex and Gender*. Staten Island, N.Y.: Gordian Press.

Hubbard, Ruth, Mary Sue Henifin, and Barbara Fried, eds. 1979. *Women Look at Biology Looking at Women: A Collection of Feminist Critiques*. Cambridge, Mass.: Schenkman Publishing.

Hull, Gloria, Patricia Bell Scott, and Barbara Smith, eds. 1982. *All the Women Are White, All the Men Are Black, But Some of Us Are Brave*. Old Westbury, N.Y.: Feminist Press.

International Fund for Agricultural Development. 1985. *IFAD Experience Relating to Rural Women, 1977–84*. Rome: IFAD, 37.

# A Cyborg Manifesto

82

Irigaray, Luce. 1977. *Ce sexe qui n'en est pas un.* Paris: Les Éditions de Minuit.

———. 1979. *Et l'une ne bouge pas sans l'autre.* Paris: Les Éditions de Minuit.

Jaggar, Alison. 1983. *Feminist Politics and Human Nature.* Totowa, N.J.: Rowman and Allenheld.

Jameson, Frederic. 1984. "Post-Modernism, or the Cultural Logic of Late Capitalism." *New Left Review* 146: 53–92.

Kahn, Douglas, and Diane Neumaier, eds. 1985. *Cultures in Contention.* Seattle: Real Comet Press.

Keller, Evelyn Fox. 1983. *A Feeling for the Organism.* San Francisco: W. H. Freeman.

———. 1985. *Reflections on Gender and Science.* New Haven, Conn.: Yale University Press.

King, Katie. 1984. "The Pleasure of Repetition and the Limits of Identification in Feminist Science Fiction: Reimaginations of the Body after the Cyborg." Presented at the California American Studies Association, Pomona.

———. 1986. "The Situation of Lesbianism as Feminism's Magical Sign: Contests for Meaning and the U.S. Women's Movement, 1968–72." *Communication* 1: 65–92.

———. 1987a. "Canons without Innocence." PhD diss., University of California, Santa Cruz.

———. 1987b. "The Passing Dreams of Choice: Audre Lorde and the Apparatus of Literary Production." Unpublished manuscript (book prospectus).

Kingston, Maxine Hong. 1976. *The Woman Warrior.* New York: Alfred A. Knopf.

Klein, Hilary. 1989. "Marxism, Psychoanalysis, and Mother Nature." *Feminist Studies* 15 (2): 255–78.

## *A Cyborg Manifesto*

Knorr-Cetina, Karin. 1981. *The Manufacture of Knowledge.* Oxford: Pergamon Press.

——, and Michael Mulkay, eds. 1983. *Science Observed: Perspectives on the Social Study of Science.* Beverly Hills, Calif.: Sage Publications.

Kramarae, Cheris, and Paula Treichler. 1985. *A Feminist Dictionary.* Boston: Pandora Press.

Kristeva, Julia. 1984. *Revolution in Poetic Language.* New York: Columbia University Press.

Latour, Bruno. 1984. *Les Microbes: guerre et paix, suivi des irréductions.* Paris: Métailié.

——, and Steve Woolgar. 1979. *Laboratory Life: The Social Construction of Scientific Facts.* Beverly Hills, Calif.: Sage Publications.

Lerner, Gerda, ed. 1973. *Black Women in White America: A Documentary History.* New York: Vintage.

Lévi-Strauss, Claude. 1973. *Tristes Tropiques.* Trans. John and Doreen Weightman. New York: Atheneum.

Lewontin, R. C., Steven Rose, and Leon J. Kamin. 1984. *Not in Our Genes: Biology, Ideology, and Human Nature.* New York: Pantheon Books.

Lorde, Audrey. 1982. *Zami: A New Spelling of My Name.* Watertown, Mass.: Persephone Press.

——. 1984. *Sister Outsider.* Trumansburg, N.Y.: Crossing Press.

Lowe, Lisa. 1986. "French Literary Orientalism: The Representation of "Others" in the Texts of Montesquieu, Flaubert, and Kristeva." PhD diss., University of California, Santa Cruz.

Mackey, Nathaniel. 1984. "Review." *Sulfur* 2: 200–205.

MacKinnon, Catharine. 1982. "Feminism, Marxism, Method, and the State: An Agenda for Theory." *Signs* 7 (3): 515–44.

——. 1987. *Feminism Unmodified: Discourses on Life and Law.* Cambridge, Mass.: Harvard University Press.

## A Cyborg Manifesto

*Many Voices, One Chant: Black Feminist Voices.* 1984. *Feminist Review* 17: special issue.

Marcuse, Herbert. 1964. *One-Dimensional Man.* Boston: Beacon Press.

Markoff, John, and Lenny Siegel. 1983. "Military Micros." Presented at Silicon Valley Research Project Conference, University of California, Santa Cruz.

Marks, Elaine, and Isabelle de Courtivron, eds. 1980. *New French Feminisms.* Amherst: University of Massachusetts Press.

McCaffery, Anne. 1969. *The Ship Who Sang.* New York: Ballantine.

———. 1978. *Dinosaur Planet.* New York: Ballantine Books.

McIntyre, Vonda. 1983. *Superluminal.* Boston: Houghton Mifflin.

———. 1978. *Dreamsnake.* New York: Dell Books.

Merchant, Carolyn. 1980. *Death of Nature: Women, Ecology, and the Scientific Revolution.* New York: Harper and Row.

Microelectronics Group. 1980. *Microelectronics: Capitalist Technology and the Working Class.* London: CSE Books.

Mohanty, Chandra Talpade. 1984. "Under Western Eyes: Feminist Scholarship and Colonial Discourse." *Boundary* 2, 3 (12/13): 333–58.

Moraga, Cherríe. 1983. *Loving in the War Years: lo que nunca paso por sus labios.* Boston: South End Press.

Moraga, Cherríe, and Gloria Anzaldúa, eds. 1981. *This Bridge Called My Back: Writings by Radical Women of Color.* Watertown, Mass.: Persephone Press.

Morgan, Robin, ed. 1984. *Sisterhood Is Global.* Garden City, N.Y.: Anchor/Doubleday.

Nash, June, and Maria Patricia Fernandez-Kelly, eds. 1983. *Women and Men and the International Division of Labor.* Albany: State University of New York Press.

Nash, Roderick. 1979. "The Exporting and Importing of Nature: Nature-

*A Cyborg Manifesto*

Appreciation as a Commodity, 1850–1980." *Perspectives in American History* 3: 517–60.

National Science Foundation. 1988. *Women and Minorities in Science and Engineering.* Washington, D.C.: NSF.

*New York Times.* 1984. "Focus of U.N. Food Day Tomorrow: Women." October 14.

O'Brien, Mary. 1981. *The Politics of Reproduction.* New York: Routledge and Kegan Paul.

Ong, Aihwa. 1987. *Spirits of Resistance and Capitalist Discipline: Factory Workers in Malaysia.* Albany: State University of New York Press.

Ong, Walter. 1982. *Orality and Literacy: The Technologizing of the Word.* New York: Methuen.

Park, Katherine, and Lorraine J. Daston. 1981. "Unnatural Conceptions: The Study of Monsters in Sixteenth- and Seventeenth-Century France and England." *Past and Present* 92: 20–54.

Perloff, Marjorie. 1984. "Dirty Language and Scramble Systems." *Sulfur* 11: 178–83.

Petschesky, Rosalind. 1981. "Abortion, Anti-feminism, and the Rise of the New Right." *Feminist Studies* 7 (2): 206–46.

Piven, Frances Fox, and Richard Coward. 1982. *The New Class War: Reagan's Attack on the Welfare State and Its Consequences.* New York: Pantheon Books.

Preston, Douglas. 1984. "Shooting in Paradise." *Natural History* 93 (12): 14–19.

Reagon, Bernice Johnson. 1983. "Coalition Politics: Turning the Century." In Smith 1983, 356–68.

Reskin, Barbara F., and Heidi Hartmann, eds. 1986. *Women's Work, Men's Work.* Washington, D.C.: National Academy of Sciences.

# A Cyborg Manifesto

86

Rich, Adrienne. 1978. *The Dream of a Common Language*. New York: W. W. Norton.

Rose, Hilary. 1983. "Hand, Brain, and Heart: A Feminist Epistemology for the Natural Sciences." *Signs* 9 (1): 73–90.

Rose, Stephen. 1986. *The American Profile Poster: Who Owns What, Who Makes How Much, Who Works, Where, and Who Lives with Whom?* New York: Pantheon Books.

Rossiter, Margaret. 1982. *Women Scientists in America*. Baltimore: Johns Hopkins University Press.

Rothschild, Joan, ed. 1983. *Machina ex Dea: Feminist Perspectives on Technology*. New York: Pergamon Press.

Russ, Joanna. 1975. *The Female Man*. New York: Bantam Books.

——. 1983a. *Adventures of Alix*. New York: Timescape.

——. 1983b. *How to Suppress Women's Writing*. Austin: University of Texas Press.

Sachs, Carolyn. 1983. *The Invisible Farmers: Women in Agricultural Production*. Totowa, N.J.: Rowman and Allenheld.

Said, Edward. 1978. *Orientalism*. New York: Pantheon Books.

Sandoval, Chela. N.d. *Yours in Struggle: Women Respond to Racism, a Report to the National Women's Studies Association*. Oakland: Center for Third World Organizing.

——. 1984. "Dis-illusionment and the Poetry of the Future: the Making of Oppositional Consciousness." PhD qualifying essay, University of California at Santa Cruz.

Schiebinger, Londa. 1987. "The History and Philosophy of Women in Science: A Review Essay." *Signs* 12 (2): 305–32.

Science Policy Research Unit. 1982. *Microelectronics and Women's Employment in Britain*. Sussex: University of Sussex.

# A Cyborg Manifesto

87

Smith, Barbara, ed. 1983. *Home Girls: A Black Feminist Anthology.* New York: Kitchen Table, Women of Color Press.

Smith, Dorothy. 1974. "Women's Perspective as a Radical Critique of Sociology." *Sociological Inquiry* 44.

———. 1979. "A Sociology of Women." In *The Prism of Sex.* Ed. J. Sherman and E. T. Beck. Madison: University of Wisconsin Press.

Sofia [Sofoulis], Zoë. 1984. "Exterminating Fetuses: Abortion, Disarmament, and the Sexo-Semiotics of Extraterrestrialism." *Diacritics* 14 (2): 47–59.

Sofoulis, Zoë. N.d. [1983?] "Lacklein." Unpublished manuscript.

Sontag, Susan. 1977. *On Photography.* New York: Dell.

Stacey, Judith. 1987. "Sexism by a Subtler Name? Postindustrial Conditions and the Postfeminist Consciousness." *Socialist Review* 96: 7–28.

Stallard, Karin, Barbara Ehrenreich, and Holly Sklar. 1983. *Poverty in the American Dream.* Boston: South End Press.

Sturgeon, Noel. 1986. "Feminism, Anarchism, and Non-Violent Direct Action Politics." PhD qualifying essay, University of California, Santa Cruz.

Sussman, Vic. 1986. "Personal Tech: Technology Lends a Hand." *Washington Post Magazine,* 9 November, 45–56.

Tiptree, James Jr. 1978a. *Star Songs of an Old Primate.* New York: Del Rey.

———. 1978b. *Up the Walls of the World.* New York: Berkeley.

Traweek, Sharon. 1988. *Beamtimes and Lifetimes: The World of High Energy Physics.* Cambridge, Mass.: Harvard University Press.

Treichler, Paula. 1987. "AIDS, Homophobia, and Biomedical Discourse: An Epidemic of Signification." *October* 43: 31–70.

## A Cyborg Manifesto

Trinh T. Minh-ha. 1986–87a. "Introduction," and "Difference: 'A Special Third World Women Issue.'" *Discourse: Journal for Theoretical Studies in Media and Culture* 8: 3–38.

——, ed. 1986–87b. *She, the Inappropriate/d Other. Discourse* 8 (Winter).

Varley, John. 1979. *Titan.* New York: Berkeley.

——. 1981. *Wizard.* New York: Berkeley.

——. 1984. *Demon.* New York: Berkeley.

Weisenbaum, Joseph. 1976. *Computer Power and Human Reason.* San Francisco: W. H. Freeman.

Wilford, John Noble. 1986. "Pilot's Helmet Helps Interpret High-Speed World." *New York Times,* July 1: 21, 24.

Wilfred, Denis. 1982. "Capital and Agriculture, a Review of Marxian Problematics." *Studies in Political Economy* 7: 127–54.

Winner, Langdon. 1977. *Autonomous Technology: Technics Out of Control as a Theme in Political Thought.* Cambridge, Mass.: MIT Press.

——. 1980. "Do Artifacts Have Politics? *Daedalus* 109 (1): 121–36.

——. 1986. *The Whale and the Reactor.* Chicago: University of Chicago Press.

Winograd, Terry, and Fernando Flores. 1986. *Understanding Computers and Cognition: A New Foundation for Design.* Norwood, N.J.: Ablex Publishing.

Wittig, Monique. 1973 [1975]. *The Lesbian Body.* Trans. David LeVay. New York: Avon.

*Women and Poverty* special issue. 1984. *Signs* 10 (2).

Wright, Susan. 1982. "Recombinant DNA: The Status of Hazards and Controls." *Environment* 24 (6): 12–20, 51–53.

——. 1986. "Recombinant DNA Technology and Its Social Transformation, 1972–82." *Osiris* (2nd series) 2: 303–60.

# A Cyborg Manifesto

89

Young, Robert M. 1979. "Interpreting the Production of Science." *New Scientist* 29 (March): 1026–28.

——, and Les Levidow, eds. 1981, 1985. *Science, Technology and the Labour Process.* 2 vols. London: CSE and Free Association Books.

Yoxen, Edward. 1983. *The Gene Business.* New York: Harper & Row.

Zimmerman, Jan, ed. 1983. *The Technology Woman: Interfacing with Tomorrow.* New York: Praeger.

*A Cyborg Manifesto*

# The Companion Species Manifesto

## DOGS, PEOPLE, AND

## SIGNIFICANT OTHERNESS

## I. EMERGENT NATURECULTURES
### From "Notes of a Sports Writer's Daughter"

Ms. Cayenne Pepper continues to colonize all my cells—a sure case of what the biologist Lynn Margulis calls symbiogenesis. I bet if you checked our DNA, you'd find some potent transfections between us. Her saliva must have the viral vectors. Surely, her darter-tongue kisses have been irresistible. Even though we share placement in the phylum of vertebrates, we inhabit not just different genera and divergent families, but altogether different orders.

How would we sort things out? Canid, hominid; pet, professor; bitch, woman; animal, human; athlete, handler. One of us has a microchip injected under her neck skin for identification; the other has a photo ID California driver's license. One of us has a written record of her ancestors for twenty generations; one of us does not know her great-grandparents' names. One of us, product of a vast genetic mixture, is called "purebred." One of us, equally product of a vast mixture, is called "white." Each of these names designates a racial discourse, and we both inherit their consequences in our flesh.

One of us is at the cusp of flaming, youthful, physical

achievement; the other is lusty but over the hill. And we play a team sport called agility on the same expropriated Native land where Cayenne's ancestors herded merino sheep. These sheep were imported from the already colonial pastoral economy of Australia to feed the California Gold Rush forty-niners. In layers of history, layers of biology, layers of naturecultures, complexity is the name of our game. We are both the freedom-hungry offspring of conquest, products of white settler colonies, leaping over hurdles and crawling through tunnels on the playing field.

I'm sure our genomes are more alike than they should be. There must be some molecular record of our touch in the codes of living that will leave traces in the world, no matter that we are each reproductively silenced females, one by age, one by surgery. Her red merle Australian Shepherd's quick and lithe tongue has swabbed the tissues of my tonsils, with all their eager immune system receptors. Who knows where my chemical receptors carried her messages, or what she took from my cellular system for distinguishing self from other and binding outside to inside?

We have had forbidden conversation; we have had oral intercourse; we are bound in telling story upon story with nothing but the facts. We are training each other in acts of communication we barely understand. We are, constitutively, companion species. We make each other up, in the flesh. Significantly other to each other, in specific difference, we signify in the flesh a nasty developmental infection called love. This love is a historical aberration and a naturalcultural legacy.

This manifesto explores two questions flowing from this aberration and legacy: (1) how might an ethics and politics committed to the flourishing of significant otherness be learned from taking dog–human relationships seriously; and (2) how might stories about dog–human worlds finally convince brain-damaged U.S. Americans, and maybe other less historically challenged people, that history matters in naturecultures?

"The Companion Species Manifesto" is a personal document, a scholarly foray into too many half-known territories, a political act of hope in a world on the edge of global war, and a work permanently in progress, in principle. I offer dog-eaten props and half-trained arguments to reshape some stories I care about a great deal, as a scholar and as a person in my time and place. The story here is mainly about dogs. Passionately engaged in these accounts, I hope to bring my readers into the kennel for life. But I hope also that even the dog phobic—or just those with their minds on higher things—will find arguments and stories that matter to the worlds we might yet live in. The practices and actors in dog worlds, human and nonhuman alike, ought to be central concerns of technoscience studies. Even closer to my heart, I want my readers to know why I consider

*The Companion Species Manifesto*

dog writing to be a branch of feminist theory, or the other way around.

This is not my first manifesto; in 1985, I published "The Cyborg Manifesto" to try to make feminist sense of the implosions of contemporary life in technoscience. Cyborgs are "cybernetic organisms," named in 1960 in the context of the space race, the Cold War, and imperialist fantasies of technohumanism built into policy and research projects. I tried to inhabit cyborgs critically, i.e., neither in celebration nor condemnation, but in a spirit of ironic appropriation for ends never envisioned by the space warriors.

Telling a story of cohabitation, coevolution, and embodied cross-species sociality, the present manifesto asks which of two cobbled-together figures—cyborgs and companion species—might more fruitfully inform livable politics and ontologies in current life worlds. These figures are hardly polar opposites. Cyborgs and companion species each bring together the human and nonhuman, the organic and technological, carbon and silicon, freedom and structure, history and myth, the rich and the poor, the state and the subject, diversity and depletion, modernity and postmodernity, and nature and culture in unexpected ways. Besides, neither a cyborg nor a companion animal pleases the pure of heart who long for better protected species boundaries and sterilization of category deviants. Nonetheless, the differences between even the most politically correct cyborg and an ordinary dog matter.

*The Companion Species Manifesto*

I appropriated cyborgs to do feminist work in Reagan's Star Wars times of the mid-1980s. By the end of the millennium, cyborgs could no longer do the work of a proper herding dog to gather up the threads needed for critical inquiry. So I go happily to the dogs to explore the birth of the kennel to help craft tools for science studies and feminist theory in the present time, when secondary Bushes threaten to replace the old growth of more livable naturecultures in the carbon budget politics of all water-based life on earth. Having worn the scarlet letters "Cyborgs for earthly survival!" long enough, I now brand myself with a slogan only Schutzhund women from dog sports could have come up with, when even a first nip can result in a death sentence: "Run fast; bite hard!"

This is a story of biopower and biosociality, as well as of technoscience. Like any good Darwinian, I tell a story of evolution. In the mode of (nucleic) acidic millennialism, I tell a tale of molecular differences, but one less rooted in Mitochondrial Eve in a neocolonial Out of Africa and more rooted in those first mitochondrial canine bitches who got in the way of man making himself yet again in the Greatest Story Ever Told. Instead, those bitches insisted on the history of companion species, a very mundane and ongoing sort of tale, one full of misunderstandings, achievements, crimes, and renewable hopes. Mine is a story told by a student of the sciences and a feminist of a certain generation who has gone to the dogs, literally. Dogs, in their historical complexity, matter here. Dogs are not an alibi for other

*The Companion Species Manifesto*

themes; dogs are fleshly material-semiotic presences in the body of technoscience. Dogs are not surrogates for theory; they are not here just to think with. They are here to live with. Partners in the crime of human evolution, they are in the garden from the get-go, wily as Coyote.

## Prehensions

Many versions of process philosophies help me walk with my dogs in this manifesto. For example, Alfred North Whitehead described "the concrete" as "a concrescence of prehensions." For him, "the concrete" meant an "actual occasion." Reality is an active verb, and the nouns all seem to be gerunds with more appendages than an octopus. Through their reaching into each other, through their "prehensions" or graspings, beings constitute each other and themselves. Beings do not preexist their relatings. "Prehensions" have consequences. The world is a knot in motion. Biological and cultural determinism are both instances of misplaced concreteness—i.e., the mistake of, first, taking provisional and local category abstractions like "nature" and "culture" for the world and, second, mistaking potent consequences to be preexisting foundations. There are no preconstituted subjects and objects, and no single sources, unitary actors, or final ends. In Judith Butler's terms, there are only "contingent foundations"; bodies that matter are the result. A bestiary of agencies, kinds of relatings, and scores of time trump the imaginings of even the most baroque cosmologists. For me, that is what *companion species* signifies.

My love of Whitehead is rooted in biology, but even more in the practice of feminist theory as I have experienced it. This feminist theory, in its refusal of typological thinking, binary dualisms, and both relativisms and universalisms of many flavors, contributes a rich array of approaches to emergence, process, historicity, difference, specificity, cohabitation, co-constitution, and contingency. Dozens of feminist writers have refused both relativism and universalism. Subjects, objects, kinds, races, species, genres, and genders are the products of their relating. None of this work is about finding sweet and nice—"feminine"—worlds and knowledges free of the ravages and productivities of power. Rather, feminist inquiry is about understanding how things work, who is in the action, what might be possible, and how worldly actors might somehow be accountable to and love each other less violently.

For example, studying Yoruba- and English-speaking mathematics elementary school classrooms in postindependence Nigeria and participating in Australian Aboriginal projects in math teaching and environmental policy, Helen Verran identifies "emergent ontologies." Verran asks "simple" questions: how can people rooted in different knowledge practices "get on together," especially when an all-too-easy cultural relativism is not an option, either politically, epistemologically, or morally?

## The Companion Species Manifesto

How can general knowledge be nurtured in postcolonial worlds committed to taking difference seriously? Answers to these questions can only be put together in emergent practices; i.e., in vulnerable, on-the-ground work that cobbles together nonharmonious agencies and ways of living that are accountable both to their disparate inherited histories and to their barely possible but absolutely necessary joint futures. For me, that is what *significant otherness* signifies.

Studying assisted reproduction practices in San Diego and then conservation science and politics in Kenya, Charis Thompson suggests the term *ontological choreographies*. The scripting of the dance of being is more than a metaphor; bodies, human and nonhuman, are taken apart and put together in processes that make self-certainty and either humanist or organicist ideology bad guides to ethics and politics, much less to personal experience.

Finally, Marilyn Strathern, drawing on decades of study of Papua New Guinean histories and politics, as well as on her investigation of English kin-reckoning habits, teaches us why conceiving of "nature" and "culture" as either polar opposites or universal categories is foolish. An ethnographer of relational categories, she has shown how to think in other topologies. Instead of opposites, we get the whole sketchpad of the modern geometrician's fevered brain with which to draw relationality. Strathern thinks in terms of "partial connections," i.e., patterns within which the players are neither wholes nor parts. I

*The Companion Species Manifesto*

call these the relations of significant otherness. I think of Strathern as an ethnographer of naturecultures; she will not mind if I invite her into the kennel for a cross-species conversation.

For feminist theorists, who and what are in the world is precisely what is at stake. This is very promising philosophical bait for training us all to understand companion species in both storied deep time, which is chemically etched in the DNA of every cell, and in recent doings, which leave more odoriferous traces. In old-fashioned terms, "The Companion Species Manifesto" is a kinship claim, one made possible by the concrescence of prehensions of many actual occasions. Companion species rest on contingent foundations.

And like the productions of a decadent gardener who can't keep good distinctions between natures and cultures straight, the shape of my kin networks looks more like a trellis or an esplanade than a tree. You can't tell up from down, and everything seems to go sidewise. Such snake-like, sidewinding traffic is one of my themes. My garden is full of snakes, full of trellises, full of indirection. Instructed by evolutionary population biologists and bioanthropologists, I know that multidirectional gene flow—multidirectional flows of bodies and values—is and has always been the name of the game of life on earth. It is certainly the way into the kennel. Whatever else humans and dogs can illustrate, it is that these large-bodied, globally distributed, ecologically opportunistic, gregariously social, mammalian co-travelers have written into their genomes a record of couplings and infectious exchanges to set the teeth of even the most committed free trader on edge. Even in the Galapagos Islands of the modern purebred dog fancy—where the effort to isolate and fragment breeding populations and deplete their heritage of diversity can look like model experiments for mimicking the natural disasters of population bottlenecks and epidemic disease—the restless exuberance of gene flow cannot be stilled. Impressed by this traffic, I risk alienating my old doppelgänger, the cyborg, in order to try to convince readers that dogs might be better guides through the thickets of technobiopolitics in the Third Millennium of the Current Era.

## Companions

In the "Cyborg Manifesto," I tried to write a surrogacy agreement, a trope, a figure for living within and honoring the skills and practices of contemporary technoculture without losing touch with the permanent war apparatus of a nonoptional, postnuclear world and its transcendent, very material lies. Cyborgs can be figures for living within contradictions, attentive to the naturecultures of mundane practices, opposed to the dire myths of self-birthing, embracing mortality as the condition for life, and alert to the emergent historical hybridities actually populating the world at all its contingent scales.

*The Companion Species Manifesto*

However, cyborg refigurations hardly exhaust the tropic work required for ontological choreography in technoscience. I have come to see cyborgs as junior siblings in the much bigger, queer family of companion species, in which reproductive bio-technopolitics are generally a surprise, sometimes even a nice surprise. I know that a U.S. middle-aged white woman with a dog playing the sport of agility is no match for the automated warriors, terrorists, and their transgenic kin in the annals of philosophical inquiry or the ethnography of naturecultures. Besides, (1) self-figuration is not my task; (2) transgenics are not the enemy; and (3) contrary to lots of dangerous and unethical projection in the Western world that makes domestic canines into furry children, dogs are not about oneself. Indeed, that is the beauty of dogs. They are not a projection, nor the realization of an intention, nor the telos of anything. They are dogs, i.e., a species in obligatory, constitutive, historical, protean relationship with human beings. The relationship is not especially nice; it is full of waste, cruelty, indifference, ignorance, and loss, as well as of joy, invention, labor, intelligence, and play. I want to learn how to narrate this cohistory and how to inherit the consequences of coevolution in natureculture.

There cannot be just one companion species; there have to be at least two to make one. It is in the syntax; it is in the flesh. Dogs are about the inescapable, contradictory story of relationships — co-constitutive relationships in which none of the partners preexists the relating, and the relating is never done once and for all. Historical specificity and contingent mutability rule all the way down, into nature and culture, into naturecultures. There is no foundation; there are only elephants supporting elephants all the way down.

Companion animals comprise only one kind of companion species, and neither category is very old in American English. In U.S. English, the term *companion animal* emerges in medical and psychosociological work in veterinary schools and related sites from the middle 1970s. This research told us that, except for those few non–dog-loving New Yorkers who obsess about unscooped dog shit in the streets, having a dog lowers one's blood pressure and ups one's chances of surviving childhood, surgery, and divorce.

Certainly, references in European languages to animals serving as companions, rather than as working or sporting dogs, predate this U.S. biomedical, technoscientific literature by centuries. Further, in China, Mexico, and elsewhere in the ancient and contemporary world, the documentary, archaeological, and oral evidence for dogs as pets, in addition to a myriad of other jobs, is strong. In the early Americas dogs assisted in hauling, hunting, and herding for various peoples. For others, dogs were food or a source of fleece. Dog people like to forget that dogs were also lethal guided weapons and instruments of terror in the European conquest of the Americas, as well as in Alexander the Great's paradigm-setting imperial travels. With combat history in Vietnam as an officer in the U.S. Marines,

Akita breeder and dog writer John Cargill reminds us that before cyborg warfare, trained dogs were among the best intelligent weapons systems. And tracking hounds terrorized slaves and prisoners, as well as rescued lost children and earthquake victims.

Listing these functions does not begin to get at the heterogeneous history of dogs in symbol and story all over the world, nor does the list of jobs tell us how dogs were treated or how they regarded their human associates. In *A History of Dogs in the Early Americas,* Marion Schwartz writes that some Native American hunting dogs went through similar rituals of preparation as did their humans, including among the Achuar of South America the ingestion of a hallucinogen. In *In the Company of Animals,* James Serpell relates that for the nineteenth-century Comanche of the Great Plains, horses were of great practical value. But horses were treated in a utilitarian way, while dogs, kept as pets, merited fond stories and warriors mourned their deaths. Some dogs were and are vermin; some were and are buried like people. Contemporary Navajo herding dogs relate to their landscape, their sheep, their people, coyotes, and dog or human strangers in historically specific ways. In cities, villages, and rural areas all over the world, many dogs live parallel lives among people, more or less tolerated, sometimes used and sometimes abused. No one term can do justice to this history.

However, the term *companion animal* enters U.S. technoculture through the post–Civil War land-grant academic institutions housing the vet schools. That is, *companion animal* has the pedigree of the mating between technoscientific expertise and late-industrial pet-keeping practices, with their democratic masses in love with their domestic partners, or at least with the nonhuman ones. Companion animals can be horses, dogs, cats, or a range of other beings willing to make the leap to the biosociality of service dogs, family members, or team members in cross-species sports. Generally speaking, one does not eat one's companion animals (or get eaten by them); and one has a hard time shaking colonialist, ethnocentric, ahistorical attitudes toward those who do (eat or get eaten).

*Species*

"Companion species" is a bigger and more heterogeneous category than companion animal, and not just because one must include such organic beings as rice, bees, tulips, and intestinal flora, all of whom make life for humans what it is—and vice versa. I want to write the keyword entry for *companion species* to insist on four tones simultaneously resonating in the linguistic, historical voice box that enables uttering this term. First, as a dutiful daughter of Darwin, I insist on the tones of the history of evolutionary biology, with its categories of populations, rates of gene flow, variation, selection, and biological species. The debates in the past 150 years about whether the category "species" denotes a real biological entity or merely figures a convenient taxonomic box sound the over- and undertones. Species is about biological kind, and scientific expertise is necessary to that kind of reality. Post-cyborg, what counts as biological kind troubles previous categories of organism. The machinic and the textual are internal to the organic and vice versa in irreversible ways.

Second, schooled by Thomas Aquinas and other Aristotelians, I remain alert to species as generic philosophical kind and category. Species is about defining difference, rooted in polyvocal fugues of doctrines of cause.

Third, my soul indelibly marked by a Catholic formation, I hear in species the doctrine of the Real Presence under both species, bread and wine, the transubstantiated signs of the flesh. Species is about the corporeal join of the material and the semiotic in ways unacceptable to the secular Protestant sensibilities of the American academy and to most versions of the human science of semiotics.

Fourth, converted by Marx and Freud and a sucker for dubious etymologies, I hear in *species* filthy lucre, specie, gold, shit, filth, wealth. In *Love's Body,* Norman O. Brown taught me about the join of Marx and Freud in shit and gold, in primitive scat and civilized metal, in specie. I met this join again in modern U.S. dog culture, with its exuberant commodity culture; its vibrant practices of love and desire; its structures that tie together the state, civil society, and the liberal individual; its mongrel technologies of purebred subject- and object-making. As I glove my hand in the plastic film — courtesy of the research empires of industrial chemistry — that protects my morning *New York Times* to pick up the microcosmic ecosystems, called scat, produced anew each day by my dogs, I find pooper scoopers quite a joke, one that lands me back in the histories of the incarnation, political economy, technoscience, and biology.

In sum, "companion species" is about a four-part composition, in which co-constitution, finitude, impurity, historicity, and complexity are what is.

"The Companion Species Manifesto" is thus about the implosion of nature and culture in the relentlessly historically specific, joint lives of dogs and people, who are bonded in significant otherness. Many are interpellated into that story, and the tale is instructive also for those who try to keep a hygienic distance. I want to convince my readers that inhabitants of technoculture become who we are in the symbiogenetic tissues of naturecultures, in story and in fact.

I take *interpellation* from the French poststructuralist and Marxist philosopher Louis Althusser's theory for how subjects are constituted from concrete individuals by being "hailed" through ideology into their subject positions in the modern state. Today, through our ideologically loaded narratives of their lives, animals "hail" us to account for the regimes in which they and we must live. We "hail" them into our constructs of nature and culture, with major consequences of life and death, health and illness, longevity and extinction. We also live with each other in the flesh in ways not exhausted by our ideologies. Stories are much bigger than ideologies. In that is our hope.

In this long philosophical introduction, I am violating a major rule of "Notes of a Sports Writer's Daughter," my doggish scribblings in honor of my sports writer father, which pepper this manifesto. The "Notes" require there to be no deviation from the animal stories themselves. Lessons have to be inextricably part of the story; it's a rule of truth as a genre for those of us—practicing and lapsed Catholics and their fellow travelers— who believe that the sign and the flesh are one.

Reporting the facts, telling a true story, I write "Notes of a Sports Writer's Daughter." A sports writer's job is, or at least was, to report the game story. I know this because as a child I sat in the press box in the AAA baseball club's Denver Bears Stadium late at night watching my father write and file his game stories. A sports writer, perhaps more than other news people, has a curious job—to tell what happened by spinning a story that is just the facts. The more vivid the prose, the better; indeed, if crafted faithfully, the more potent the tropes, the truer the story. My father did not want to have a sports column, a more prestigious activity in the newspaper business. He wanted to write the game stories, to stay close to the action, to tell it like it is, not to look for the scandals and the angles for the metastory, the column. My father's faith was in the game, where fact and story cohabit.

*The Companion Species Manifesto*

I grew up in the bosom of two major institutions that counter the modernist belief in the no-fault divorce, based on irrevocable differences, of story and fact. Both of these institutions—the Church and the Press—are famously corrupt, famously scorned (if constantly used) by Science, and nonetheless indispensable in cultivating a people's insatiable hunger for truth. Sign and flesh; story and fact. In my natal house, the generative partners could not separate. They were, in down-and-dirty dog talk, tied. No wonder culture and nature imploded for me as an adult. And nowhere did that implosion have more force than in living the relationship and speaking the verb that passes as a noun: companion species. Is this what John meant when he said, "The Word was made flesh"? In the bottom of the ninth inning, the Bears down by two runs, with three on, two out, and two strikes, with the time deadline for filing the story five minutes away?

I also grew up in the house of Science and learned at around the time my breast buds erupted about how many underground passages there are connecting the Estates and how many couplings keep sign and flesh, story and fact, together in the palaces of positive knowledge, falsifiable hypothesis, and synthesizing theory. Because my science was biology, I learned early that accounting for evolution, development, cellular function, genome complexity, the molding of form across time, behavioral ecology, systems communication, cognition—in short, accounting for anything worthy of the name of biology—was not so different from getting a game story filed or living with the conundrums of the incarnation. To do biology with any kind of fidelity, the practitioner *must* tell a story, *must* get the facts, and *must* have the heart to stay hungry for the truth and to abandon a favorite story, a favorite fact, shown to be somehow off the mark. The practitioner must also have the heart to stay with a story through thick and thin, to inherit its discordant resonances, to live its contradictions, when that story gets at a truth about life that matters. Isn't that kind of fidelity what has made the science of evolutionary biology flourish and feed my people's corporeal hunger for knowledge over the past hundred and fifty years?

Etymologically, facts refer to performance, action, deeds done—feats, in short. A fact is a past participle, a thing done, over, fixed, shown, performed, accomplished. Facts have made the deadline for getting into the next edition of the paper. Fiction, etymologically, is very close but differs by part of speech and tense. Like facts, fiction refers to action, but fiction is about the act of fashioning, forming, inventing, as well as feigning or feinting. Drawn from a present participle, fiction is in process and still at stake, not finished, still prone to falling afoul of facts, but also liable to showing something we do not yet know to be true but will know. Living with animals, inhabiting their/our stories, trying to tell the truth about relationship, cohabiting an active history: that is the work of companion species, for whom "the relation" is the smallest possible unit of analysis.

So, I file dog stories for a living these days. All stories traffic

*The Companion Species Manifesto*

in tropes, i.e., figures of speech necessary to say anything at all. Trope (Greek: *tropós*) means swerving or tripping. All language swerves and trips; there is never direct meaning, only the dogmatic think that trope-free communication is our province. My favorite trope for dog tales is "metaplasm." Metaplasm means a change in a word, for example, by adding, omitting, inverting, or transposing its letters, syllables, or sounds. The term is from the Greek *metaplasmos,* meaning remodeling or remolding. *Metaplasm* is a generic term for almost any kind of alteration in a word, intentional or unintentional. I use *metaplasm* to mean the remodeling of dog and human flesh, remolding the codes of life, in the history of companion-species relating.

Compare and contrast *protoplasm, cytoplasm, neoplasm,* and *germplasm.* There is a biological taste to *metaplasm*—just what I like in words about words. Flesh and signifier, bodies and words, stories and worlds: these are joined in naturecultures. *Metaplasm* can signify a mistake, a stumbling, a troping that makes a fleshly difference. For example, a substitution in a string of bases in a nucleic acid can be a metaplasm, changing the meaning of a gene and altering the course of a life. Or, a remolded practice among dog breeders, such as doing more outcrosses and fewer close-line breedings, could result from changed meanings of a word like *population* or *diversity.* Inverting meanings; transposing the body of communication; remolding, remodeling; swervings that tell the truth: I tell stories about stories, all the way down. Woof.

### The Companion Species Manifesto

Implicitly, this manifesto is about more than the relation of dogs and people. Dogs and people figure a universe. Clearly, cyborgs—with their historical congealings of the machinic and the organic in the codes of information, where boundaries are less about skin than about statistically defined densities of signal and noise—fit within the taxon of companion species. That is to say, cyborgs raise all the questions of histories, politics, and ethics that dogs require. Care, flourishing, differences in power, scales of time—these matter for cyborgs. For example, what kind of temporal scale-making could shape labor systems, investment strategies, and consumption patterns in which the generation time of information machines became compatible with the generation times of human, animal, and plant communities and ecosystems? What is the right kind of pooper scooper for a computer or a personal digital assistant? At the least, we know it is not an electronics dump in Mexico or India, where human scavengers get paid less than nothing for processing the ecologically toxic waste of the well informed.

Art and engineering are natural sibling practices for engaging companion species. Thus, human–landscape couplings fit snugly within the category of companion species, evoking all the questions about the histories and relatings that weld the souls of dogs and their humans. The Scots sculptor Andrew Goldsworthy understands this well. Scales and flows of time through the flesh of plants, earth, sea, ice, and stone consume Goldsworthy. For him, the history of the land is living; and that

*The Companion Species Manifesto*

history is composed out of the polyform relatings of people, animals, soil, water, and rocks. He works at scales of sculpted ice crystals interlaced with twigs, layered rock cones the size of a man built in the surging intertidal zones of the shore, and stone walls across long stretches of countryside. He has an engineer's and an artist's knowledge of forces like gravity and friction. His sculptures endure sometimes for seconds, sometimes for decades; but mortality and change are never out of consciousness. Process and dissolution—and agencies both human and nonhuman, as well as animate and inanimate—are his partners and materials, not just his themes.

In the 1990s, Goldsworthy did a work called *Arch*. He and writer David Craig traced an ancient drover's sheep route from Scottish pastures to an English market town. Photographing as they went, they assembled and disassembled a self-supporting red sandstone arch across places marking the past and present history of animals, people, and land. The missing trees and cottars, the story of the enclosures and rising wool markets, the fraught ties between England and Scotland over centuries, the conditions of possibility of the Scottish working sheepdog and hired shepherd, the sheep eating and walking to shearing and slaughter—these are memorialized in the moving rock arch tying together geography, history, and natural history.

The collie implicit in Goldsworthy's *Arch* is less about "Lassie come home" than "cottar get out." That is one condition of possibility of the immensely popular late-twentieth-

*Border Collie hell. The famous English sheepdog trialer and author Thomas Longton is the handler in this photograph.*

century British TV show about the brilliant working sheepdogs, the Border Collies of Scotland. Shaped genetically by competitive sheep trialing since the late nineteenth century, this breed has made that sport justly famous on several continents. This is the same breed of dog that dominates the sport of agility in my life. It is also the breed that is thrown away in large numbers to be rescued by dedicated volunteers or killed in animal shelters because people watching those famous TV shows about those talented dogs want to buy one on the pet market, which mushrooms to fill the demand. The impulse buyers quickly find themselves with a serious dog whom they cannot satisfy with the work the Border Collie needs. And where is the labor of the

*The Companion Species Manifesto*

hired shepherds and of the food-and-fiber-producing sheep in this story? In how many ways do we inherit in the flesh the turbulent history of modern capitalism?

How to live ethically in these mortal, finite flows that are about heterogeneous relationship—and not about "man"—is an implicit question in Goldsworthy's art. His art is relentlessly attuned to specific human inhabitations of the land, but it is neither humanist nor naturalist art. It is the art of naturecultures. The relation is the smallest unit of analysis, and the relation is about significant otherness at every scale. That is the ethic, or perhaps better, mode of attention, with which we must approach the long cohabitings of people and dogs.

So, in "The Companion Species Manifesto," I want to tell stories about relating in significant otherness, through which the partners come to be who we are in flesh and sign. The following shaggy dog stories about evolution, love, training, and kinds or breeds help me think about living well together with the host of species with whom human beings emerge on this planet at every scale of time, body, and space. The accounts I offer are idiosyncratic and indicative rather than systematic, tendentious more than judicious, and rooted in contingent foundations rather than clear and distinct premises. Dogs are my story here, but they are only one player in the large world of companion species. Parts don't add up to wholes in this manifesto—or in life in naturecultures. Instead, I am looking for Marilyn

Strathern's "partial connections," which are about the counter-intuitive geometries and incongruent translations necessary to getting on together, where the god-tricks of self-certainty and deathless communion are not an option.

## II. EVOLUTION STORIES

Everyone I know likes stories about the origin of dogs. Over-stuffed with significance for their avid consumers, these stories are the stuff of high romance and sober science all mixed up together. Histories of human migrations and exchanges, the nature of technology, the meanings of wildness, and the relations of colonizers and colonized suffuse these stories. Matters like judging whether my dog loves me, sorting out scales of intelligence among animals and between animals and humans, and deciding whether humans are the masters or the duped can hang on the outcome of a sober scientific report. Evaluating the decadence or the progressiveness of breeds, judging whether dog behavior is the stuff of genes or rearing, adjudicating between the claims of old-fashioned anatomists and archaeologists or new-fangled molecular wizards, establishing origins in the New or Old World, figuring the ancestor of pooches as a noble hunting wolf persisting in modern endangered species or a cringing scavenger mirrored in mere village dogs, looking to one or many

*The Companion Species Manifesto*

canine Eves surviving in their mitochondrial DNA or perhaps to a canine Adam through his Y-chromosome legacies—all these and more are understood to be at stake.

The day I wrote this section of "The Companion Species Manifesto," news broke on the major networks from PBS to CNN about three papers in *Science* magazine on dog evolution and the history of domestication. Within minutes, numerous email lists in dogland were abuzz with discussion about the implications of the research. Website addresses flew across continents bringing the news to the cyborg world, while the merely literate followed the story in the daily papers of New York, Tokyo, Paris, or Johannesburg. What is going on in this florid consumption of scientific origin stories, and how can these accounts help me understand the relation that is companion species?

Explanations of primate, and especially hominid, evolution might be the most notorious cock-fighting arena in contemporary life sciences; but the field of canine evolution is hardly lacking in impressive dog fights among the human scientists and popular writers. No account of the appearance of dogs on earth goes unchallenged, and none goes unappropriated by its partisans. In both popular and professional dog worlds what is at stake is twofold: (1) the relation between what counts as nature and what counts as culture in Western discourse and its cousins, and (2) the correlated issue of who and what counts as an actor. These things matter for political, ethical, and emotional action in technoculture. A partisan in the world of dog evolutionary stories, I look for ways of getting coevolution and co-constitution without stripping the story of its brutalities as well as multiform beauties.

Dogs are said to be the first domestic animals, displacing pigs for primal honors. Humanist technophiliacs depict domestication as the paradigmatic act of masculine, single-parent self-birthing, whereby man makes himself repetitively as he invents (creates) his tools. The domestic animal is the epoch-changing tool, realizing human intention in the flesh, in a dogsbody version of onanism. Man took the (free) wolf and made the (servant) dog and so made civilization possible. Mongrelized Hegel and Freud in the kennel? Let the dog stand for all domestic plant and animal species, subjected to human intent in stories of escalating progress or destruction, according to taste. Deep ecologists love to believe these stories in order to hate them in the name of Wilderness before the Fall into Culture, just as humanists believe them in order to fend off biological encroachments on culture.

These conventional accounts have been thoroughly reworked in recent years, when distributed everything is the name of the game all over, including in the kennel. Even though I know they are faddish, I like these metaplasmic, remodeled versions that give dogs (and other species) the first moves in domestication and then choreograph an unending dance of distributed and heterogeneous agencies. Besides being faddish, the newer stories, I think, have a better chance of being true, and they certainly have a better chance of teaching us to pay attention to significant otherness as something other than a reflection of one's intentions.

Studies of dog mitochondrial DNA as molecular clocks have indicated emergence of dogs earlier than previously thought possible. Work out of Carles Vilá and Robert Wayne's lab in 1997 argued for divergence of dogs from wolves as long as 150,000 years ago—that is, at the origin of *Homo sapiens sapiens*. That date, unsupported by fossil or archaeological evidence, has given way in subsequent DNA studies to somewhere from 50,000 to 15,000 years ago, with the scientists favoring the more recent date because it allows synthesis of all the available kinds of evidence. In that case, it looks like dogs emerged first somewhere in East Asia over a fairly brief time in a distributed pocket of events and then spread fast over the whole earth, going wherever humans went.

Many interpreters argue that the most likely scenario has wolf-wannabe dogs first taking advantage of the calorie bonanzas provided by humans' waste dumps. By their opportunistic moves, those emergent dogs would be behaviorally and ultimately genetically adapted for reduced-tolerance distances, less hair-trigger flight, puppy developmental timing with longer windows for cross-species socialization, and more confident parallel occupation of areas also occupied by dangerous humans. Studies of Russian fur foxes selected over many generations for differential tameness show many of the morphological and behavioral traits associated with domestication. They might model the emergence of a kind of proto-"village dog," genetically close to wolves, as all dogs remain, but behaviorally quite different and receptive to human attempts to further the domestication process. Both by deliberate control of dogs' reproduction (e.g., killing unwanted puppies or feeding some bitches and not others) and by unintended but nonetheless potent consequences, humans could have contributed to shaping the many kinds of dogs that appeared early in the story. Human life ways changed significantly in association with dogs. Flexibility and opportunism are the name of the game for both species, who shape each other throughout the still ongoing story of coevolution.

Scholars use versions of this story to question sharp divisions of nature and culture in order to shape a more generative discourse for technoculture. Darcy Morey, a canine paleobiologist and archaeologist, believes that the distinction between artificial and natural selection is empty because all the way down the story is about differential reproduction. Morey deemphasizes intentions and foregrounds behavioral ecology. Ed Russell, an environmental historian, historian of technology, and science studies scholar, argues that the evolution of dog breeds is a chapter in the history of biotechnology. He emphasizes human agencies and regards organisms as engineered technologies, but in a way that has the dogs active, as well as in a way to foreground the ongoing coevolution of human cultures and dogs.

*The Companion Species Manifesto*

The science writer Stephen Budiansky insists that domestication in general, including the domestication of dogs, is a successful evolutionary strategy benefiting humans and their associated species alike. Examples can be multiplied.

These accounts taken together require reevaluating the meanings of domestication and coevolution. Domestication is an emergent process of cohabiting, involving agencies of many sorts and stories that do not lend themselves to yet one more version of the Fall or to an assured outcome for anybody. Cohabiting does not mean fuzzy and touchy-feely. Companion species are not companionate mates ready for early-twentieth-century Greenwich Village anarchist discussions. Relationship is multiform, at stake, unfinished, consequential.

Coevolution has to be defined more broadly than biologists habitually do. Certainly, the mutual adaptation of visible morphologies like flower sexual structures and the organs of their pollinating insects is coevolution. But it is a mistake to see the alterations of dogs' bodies and minds as biological and the changes in human bodies and lives, for example in the emergence of herding or agricultural societies, as cultural, and so not about coevolution. At the least, I suspect that human genomes contain a considerable molecular record of the pathogens of their companion species, including dogs. Immune systems are not a minor part of naturecultures; they determine where organisms, including people, can live and with whom. The history of the flu is unimaginable without the concept of the coevolution of humans, pigs, fowl, and viruses.

*The Companion Species Manifesto*

But disease can't be the whole biosocial story. Some commentators think that even something as fundamental as the hypertrophied human biological capacity for speech emerged in consequence of associated dogs' taking on scent and sound alert jobs and so freeing the human face, throat, and brain for chat. I am skeptical of that account; but I am sure that once we reduce our own fight-or-flight reaction to emergent naturecultures, and stop seeing only biological reductionism or cultural uniqueness, both people and animals will look different.

I am heartened by recent ideas in ecological developmental biology, or "eco-devo" in the terms of developmental biologist and historian of science Scott Gilbert. Developmental triggers and timing are the key objects for this young science made possible by new molecular techniques and by discursive resources from many disciplines. Differential, context-specific plasticities are the rule, sometimes genetically assimilated and sometimes not. How organisms integrate environmental and genetic information at all levels, from the very small to the very large, determines what they become. There is no time or place at which genetics ends and environment begins, and genetic determinism is at best a local word for narrow ecological developmental plasticities.

The big, wide world is full of bumptious life. For example, Margaret McFall-Ngai has shown that the light-sensing organs of the squid *Euprymna scolopes* develop normally only if the embryo has been colonized by luminescent *Vibrio* bacteria. Similarly, human gut tissue cannot develop normally without colonization by its bacterial flora. The diversity of earth's animal forms emerged in the oceans' salty bacterial soup. All stages of the life histories of evolving animals had to adapt to eager bacteria colonizing them inside and out. Developmental patterns of complex life forms are likely to show the history of these adaptations, once scientists figure out how to look for the evidence. Earth's beings are prehensile, opportunistic, ready to yoke unlikely partners into something new, something symbiogenetic. Co-constitutive companion species and coevolution are the rule, not the exception. These arguments are tropic for my manifesto, but flesh and figure are not far apart. Tropes are what make us want to look and need to listen for surprises that get us out of inherited boxes.

## III. LOVE STORIES

Commonly in the United States, dogs are credited with the capacity for "unconditional love." According to this belief, people, burdened with misrecognition, contradiction, and complexity in their relations with other humans, find solace in unconditional love from their dogs. In turn, people love their dogs as children. In my opinion, both of these beliefs are not only based on mistakes, if not lies, but also they are in themselves abusive—to dogs and to humans. A cursory glance shows that dogs and humans have always had a vast range of ways of relating. But even among the pet-keeping folk of contemporary consumer cultures, or maybe especially among these people, belief in "unconditional love" is pernicious. If the idea that man makes himself by realizing his intentions in his tools, such as domestic animals (dogs) and computers (cyborgs), is evidence of a neurosis that I call humanist technophiliac narcissism, then the superficially opposed idea that dogs restore human beings' souls by their unconditional love might be the neurosis of caninophiliac narcissism. Because I find the love of and between historically situated dogs and humans precious, dissenting from the discourse of unconditional love matters.

J. R. Ackerley's quirky masterpiece *My Dog Tulip* (first privately printed in England in 1956), about a relationship between the writer and his "Alsatian" bitch in the 1940s and 1950s, gives me a way to think through my dissent. History flickers in the reader's peripheral vision from the start of this great love story. After two world wars, in one of those niggling examples of denial and substitution that allow us to go about our lives, a German Shepherd Dog in England was called an Alsatian. Tulip (Queenie, in real life) was the great love of Ackerley's life. An important novelist, famous homosexual, and splendid writer, Ackerley honored that love from the start by recognizing his impossible task—to wit, first, somehow to learn what *this* dog needed and desired and, second, to move heaven and earth to make sure she got it.

In Tulip, rescued from her first home, Ackerley hardly had

*The Companion Species Manifesto*

his ideal love object. He also suspected he was not her idea of the loved one. The saga that followed was not about unconditional love, but about seeking to inhabit an intersubjective world that is about meeting the other in all the fleshly detail of a mortal relationship. Barbara Smuts, the behavioral bioanthropologist who writes courageously about intersubjectivity and friendship with and among animals, would approve. No behavioral biologist, but attuned to the sexology of his culture, Ackerley comically and movingly sets out to find an adequate sexual partner for Tulip in her periodic heats.

The Dutch environmental feminist Barbara Noske, who also calls our attention to the scandal of the meat-producing "animal-industrial complex," suggests thinking about animals as "other worlds" in a science fictional sense. In his unswerving dedication to his dog's significant otherness, Ackerley would have understood. Tulip mattered, and that changed them both. He also mattered to her, in ways that could only be read with the tripping proper to any semiotic practice, linguistic or not. The misrecognitions were as important as the fleeting moments of getting things right. Ackerley's story is full of the fleshly, meaning-making details of worldly, face-to-face love. Receiving unconditional love from another is a rarely excusable neurotic fantasy; striving to fulfill the messy conditions of being in love is quite another matter. The permanent search for knowledge of the intimate other, and the inevitable comic and tragic mistakes in that quest, commands my respect, whether the other is animal or human, or indeed, inanimate. Ackerley's relationship with Tulip earned the name of love.

I have benefited from the mentoring of several lifelong dog people. These people use the word *love* sparingly because they loathe how dogs get taken for cuddly, furry, child-like dependents. For example, Linda Weisser has been a breeder for more than thirty years of Great Pyrenees livestock guardian dogs, a health activist in the breed, and a teacher on all aspects of these dogs' care, behavior, history, and well-being. Her sense of responsibility to the dogs and to the people who have them is stunning. Weisser emphasizes love of a *kind* of dog, of a breed, and talks about what needs to be done if people care about these dogs as a whole, and not just about one's own dogs. Without wincing, she recommends killing an aggressive rescue dog or any dog who has bitten a child; doing so could mean saving the reputation of the breed and the lives of other dogs, not to mention children. The "whole dog" for her is both a kind and an individual. This love leads her and others with very modest middle-class means to scientific and medical self-education, public action, mentoring, and major commitments of time and resources.

Weisser also talks about the special "dog of her heart"—a bitch who lived with her many years ago and who still stirs her. She writes in acid lyricism about a current dog who arrived at her house at eighteen months of age and snarled for three days, but who now accepts cookies from her nine-year-old granddaughter, allows the child to take away both food and toys, and tolerantly rules the household's younger bitches. "I love this bitch beyond words. She is smart and proud and alpha, and if a snarl here and there is the price I pay for her in my life, so be it" (Great Pyrenees Discussion List, September 29, 2002). Weisser plainly treasures these feelings and these relationships. She is quick to insist that at root her love is about "the deep pleasure, even joy, of sharing life with a different being, one whose thoughts, feelings, reactions, and probably survival needs are different from ours. And somehow in order for all the species in this 'band' to thrive, we have to learn to understand and respect those things" (Great Pyrenees Discussion List, November 14, 2001).

To regard a dog as a furry child, even metaphorically, demeans dogs and children—and sets up children to be bitten and dogs to be killed. In 2001, Weisser had eleven dogs and five cats in residence. All of her adult life, she has owned, bred, and shown dogs; and she raised three human children and carried on a full civic, political life as a subtle left feminist. Sharing human language with her children, friends, and comrades is irreplaceable. "While my dogs can love me (I think), I have never had an interesting political conversation with any of them. On the other hand, while my children can talk, they lack the true 'animal' sense that allows me to touch, however briefly, the 'being' of another species so different from my own with all the awe-inspiring reality that brings me" (Great Pyrenees Discussion List, November 14, 2001).

Loving dogs the way Weisser means it is not incompatible with a pet relationship; indeed, pet relationships can and do frequently nurture this sort of love. Being a pet seems to me to be a demanding job for a dog, requiring self-control and canine emotional and cognitive skills matching those of good working dogs. Very many pets and pet people deserve respect. Further, play between humans and pets, as well as simply spending time peaceably hanging out together, brings joy to all the participants. Surely that is one important meaning of companion species. Nonetheless, the status of pet puts a dog at special risk in societies like the one I live in—the risk of abandonment when human affection wanes, when people's convenience takes precedence, or when the dog fails to deliver on the fantasy of unconditional love.

Many of the serious dog people I have met doing my research emphasize the importance to dogs of jobs that leave them less vulnerable to human consumerist whims. Weisser knows many livestock people whose guardian dogs are respected for the work they do. Some are loved and some are not, but their value does not depend on an economy of affection. In particular, the dogs' value—and life—does not depend on the humans' perception that the dogs love them. Rather, the dog has to do his or her job, and, as Weisser says, the rest is gravy.

*The Companion Species Manifesto*

*Marco Harding and Willem de Kooning, Susan Caudill's pet Great Pyrenees, bred by and co-owned with Linda Weisser. Photograph by the author.*

Donald McCaig, the astute Border Collie writer and sheep-dog trialer, concurs. His novels *Nop's Hope* and *Nop's Trial* are a superb introduction to potent relationships between working sheepdogs and their people. McCaig notes that working sheep-dogs, as a category, fall "somewhere between 'livestock' and 'co-worker'" (Canine Genetics Discussion List, November 30, 2000). A consequence of that status is that the dog's judgment may sometimes be better than the human's on the job. Respect and trust, not love, are the critical demands of a good working relationship between these dogs and humans. The dog's life depends more on skill—and on a rural economy that does not collapse—and less on a problematic fantasy.

In his zeal to foreground the need to breed, train, and work to sustain the precious herding abilities of the breed he best knows and most cares about, I think McCaig sometimes devalues and mis-describes both pet and sport performance relationships in dogland. I also suspect that his dealings with his dogs might properly be called *love* if that word were not so corrupted by our culture's infantilization of dogs and the refusal to honor differ-ence. Dog naturecultures need his insistence on the functional dog preserved only by deliberate work-related practices, in-cluding breeding and economically viable jobs. We need Weis-ser's and McCaig's knowledge of the job of a kind of dog, the whole dog, the specificity of dogs. Otherwise, love kills, uncon-ditionally, both kinds and individuals.

*From "Notes of a Sports Writer's Daughter"*

Marco, my godson, is Cayenne's god kid; she is his god dog. We are a fictive kin group in training. Perhaps our family coat of arms would take its motto from the Berkeley canine literary, politics, and arts magazine that is modeled after the *Barb*; namely, the *Bark,* whose masthead reads "Dog is my co-pilot." When Cayenne was twelve weeks old and Marco six years old, my husband, Rusten, and I gave him puppy-training lessons for Christmas. With Cayenne in her crate in the car, I would pick Marco up from school on Tuesdays, drive to Burger King for a planet-sustaining health food dinner of burgers, Coke, and fries, and then head to the Santa Cruz SPCA for our lesson. Like many of her breed, Cayenne was a smart and willing youngster, a natural to obedience games. Like many of his generation raised on high-speed visual special effects and automated cyborg toys, Marco was a bright and motivated trainer, a natural to control games.

Cayenne learned cues fast, and so she quickly plopped her bum on the ground in response to a "sit" command. Besides, she practiced at home with me. Entranced, Marco at first treated her like a microchip-implanted truck for which he held the remote controls. He punched an imaginary button; his puppy magically fulfilled the intentions of his omnipotent, remote will. God was threatening to become our co-pilot. I, an obsessive adult who came of age in the communes of the late 1960s, was committed to ideals of intersubjectivity and mutuality in all things, certainly including dog and boy training. The illusion of mutual attention and communication would be better than nothing, but I really wanted more than that. Besides, here I was the only adult of either species present. Intersubjectivity does not mean "equality," a literally deadly game in dogland; but it does mean paying attention to the conjoined dance of face-to-face significant otherness. In addition, control freak that I am, I got to call the shots, at least on Tuesday nights.

Marco was at the same time taking karate lessons, and he was profoundly in love with his karate master. This fine man understood the children's love of drama, ritual, and costume, as well as the mental-spiritual-bodily discipline of his martial art. *Respect* was the word and the act that Marco ecstatically told me about from his lessons. He swooned at the chance to collect his small, robed self into the prescribed posture and bow formally to his master or his partner before performing a form. Calming his turbulent first-grade self and meeting the eyes of his teacher or his partner in preparation for demanding, stylized action thrilled him. Hey, was I going to let an opportunity like that go unused in my pursuit of companion species flourishing?

"Marco," I said, "Cayenne is not a cyborg truck; she is your partner in a martial art called obedience. You are the older partner and the master here. You have learned how to perform respect with your body and your eyes. Your job is to teach the form to Cayenne. Until you can find a way to teach her how to collect her galloping puppy self calmly and to hold still and look you in the eyes, you cannot let her perform the 'sit' command." It would not be enough for her just to sit on cue and for him to "click and treat." That would be necessary, certainly, but the order was wrong. First, these two youngsters had to learn to notice each other. They had to be in the same game. It is my belief that Marco began to emerge as a dog trainer over the next six weeks. It is also my belief that as he learned to show her the corporeal posture of cross-species respect, she and he became significant others to each other.

Two years later out of the kitchen window I glimpsed Marco in the backyard doing twelve weave poles with Cayenne when nobody else was present. The weave poles are one of the most difficult agility objects to teach and to perform. I think Cayenne and Marco's fast, beautiful weave poles were worthy of his karate master.

### Positive Bondage

In 2002, the consummate agility competitor and teacher Susan Garrett authored a widely acclaimed training pamphlet called *Ruff Love,* published by the dog agility-oriented company Clean Run Productions. Informed by behaviorist learning theory and the resultant popular positive training methods that have mushroomed in dogland in the past twenty years, the booklet instructs any dog person who wants a closer, more responsive training relationship with her or his dog. Problems like a dog's not coming when called or inappropriate aggression are surely in view; but, more, Garrett works to inculcate attitudes informed by biobehavioral research and to put effective tools in the hands of her agility students. She aims to show how to craft a relationship of energetic attention that would be rewarding to the dogs and the humans. Non-optional, spontaneous, oriented enthusiasm is to be the accomplishment of the previously most lax, distracted dog. I have the strong sense that Marco has been the subject of a similar pedagogy at his progressive elementary school. The rules are simple in principle and cunningly demanding in practice; to wit, mark the desired behavior with an instantaneous signal and then get a reward delivered within the time window appropriate to the species in question. The mantra of popular positive training, "click and treat," is only the tip of a vast post-"discipline and punish" iceberg.

Emphatically, as the back of Garrett's tract proclaims in a cartoon, *positive* does not mean permissive. Indeed, I have never read a dog-training manual more committed to near total control in the interests of fulfilling human intentions, in this case, peak performance in a demanding, dual species, competitive sport. That kind of performance can only come from a team that is highly motivated, not working under compulsion, but knowing the energy of each other and trusting the honesty and coherence of directional postures and responsive movements.

### The Companion Species Manifesto

Garrett's method is exacting, philosophically and practically. The human partner must set things up so that the dog sees the clumsy biped as the source of all good things. Opportunities for the dog to get rewards in any other way must be eliminated as far as possible for the duration of the training program, typically a few months. The romantic might quail in the face of requirements to keep one's dog in a crate or tied to oneself by a loose leash. Forbidden to the pooch are the pleasures of romping at will with other dogs, rushing after a teasing squirrel, or clambering onto the couch—unless and until such pleasures are granted for exhibiting self-control and responsiveness to the human's commands at a near 100 percent frequency. The human must keep detailed records of the *actual* correct response rate of the dog for each task, rather than tell tales about the heights of genius one's own dog must surely have reached. A dishonest human is in deep trouble in the world of ruff love.

The compensations for the dog are legion. Where else can a canine count on several focused training sessions a day, each designed so that the dog does not make mistakes but instead gets rewarded by the rapid delivery of treats, toys, and liberties, all carefully calibrated to evoke and sustain maximum motivation from the particular, individually known pupil? Where else in dogland do training practices lead to a dog who has learned to learn and who eagerly offers novel "behaviors" that might become incorporated into sports or living routines, instead of morosely complying (or not) with poorly understood compulsions?

Garrett directs the human to make careful lists of what the dog actually likes; and she instructs people how to play with their companions in a way *the dogs* enjoy, instead of shutting dogs down by mechanical human ball tosses or intimidating overexuberance. Besides all that, the human *must* actually enjoy playing in doggishly appropriate ways, or they will be found out. Each game in Garrett's book might be geared to build success according to human goals, but unless the game engages the dog, it is worthless.

In short, the major demand on the human is precisely what most of us don't even know we don't know how to do — to wit, how to see who the dogs are and hear what they are telling us, not in bloodless abstraction, but in one-on-one relationship, in otherness-in-connection.

There is no room for romanticism about the wild heart of the natural dog or illusions of social equality across the class Mammalia in Garrett's practice and pedagogy, but there is large space for disciplined attention and honest achievement. Psychological and physical violence has no part in this training drama; technologies of behavioral management have a starring role. I have made enough well-intentioned training mistakes—some of them painful to my dogs and some of them dangerous to people and other dogs, not to mention worthless for succeeding in agility—to pay attention to Garrett. Scientifically informed, empirically grounded practice matters; and learning theory is not empty cant, even if it is still a severely limited discourse and a rough instrument. Nonetheless, I am enough of a cultural critic to be unable to still the roaring ideologies of tough love in high-pressure, success-oriented, individualist America. Twentieth-century Taylorite principles of scientific management and the personnel management sciences of corporate America have found a safe crate around the postmodern agility field. I am enough of a historian of science to be unable to ignore the easily inflated, historically decontextualized, and overly generalized claims of method and expertise in positive training discourse.

Still, I lend my well-thumbed copy of *Ruff Love* to friends, and I keep my clicker and liver treats in my pocket. More to the point, Garrett makes me own up to the stunning capacity that dog people like me have to lie to ourselves about the conflicting fantasies we project onto our dogs in our inconsistent training and dishonest evaluations of what is actually happening. Her pedagogy of positive bondage makes a serious, historically specific kind of freedom *for dogs* possible, i.e., the freedom to live safely in multispecies, urban and suburban environments with very little physical restraint and no corporal punishment while getting to play a demanding sport with every evidence of self-actualizing motivation. In dogland, I am learning what my college teachers meant in their seminars on freedom and authority. I think my dogs rather like ruff tough love. Marco remains more skeptical.

## Harsh Beauty

Vicki Hearne—the famous companion animal trainer, lover of maligned dogs like American Staffordshire Terriers and Airedales, and language philosopher—is at first glance the opposite of Susan Garrett. Hearne, who died in 2001, remains a sharp thorn in the paw for the adherents of positive training methods. To the horror of many professional trainers and ordinary dog folk, including myself, who have undergone a near-religious conversion from the military-style Koehler dog-training methods, not so fondly remembered for corrections like leash jerks and ear pinches, to the joys of rapidly delivering liver cookies under the approving eye of behaviorist learning theorists, Hearne did not turn from the old path and embrace the new. Her disdain for clicker training could be searing, exceeded only by her fierce opposition to animal rights discourse. I cringe under her ear pinching of my newfound training practices and rejoice in her alpha roll of animal rights ideologies. The coherence and power of Hearne's critique of both the clicker addicted and the rights besotted, however, command my respect and alert me to a kinship link. Hearne and Garrett are blood sisters under the skin.

The key to this close line breeding is their focused attention to what the dogs are telling them, and so demanding of them. Amazing grace, these thinkers attend to the dogs, in all these canines' situated complexity and particularity, as the unconditional demand of their relational practice. There is no doubt that behaviorist trainers and Hearne have important differences over methods, some of which could be resolved by empirical research and some of which are embedded in personal talent and cross-species charisma or in the incommensurable tacit knowledges of diverse communities of practice. Some of the differences also probably reside in human pigheadedness and canine opportunism. But "method" is not what matters most among companion species; "communication" across irreducible difference is what matters. Situated partial connection is what matters; the resultant dogs and humans emerge together in that game of cat's cradle. Respect is the name of the game. Good trainers practice the discipline of companion species relating under the sign of significant otherness.

Hearne's best-known book about communication between companion animals and human beings, *Adam's Task,* is ill titled. The book is about two-way conversation, not about naming. Adam had it easy in his categorical labor. He didn't have to worry about back talk; and God, not a dog, made him who he was, in His own image, no less. To make matters harder, Hearne has to worry about conversation when human language isn't the medium, but not for reasons most linguists or language philosophers would give. Hearne likes trainers' using ordinary language in their work; that use turns out to be important to understanding what the dogs might be telling her, but not because the dogs are speaking furry humanese. She adamantly defends lots of so-called anthropomorphism, and no one more eloquently makes the case for the intention-laden, consciousness-ascribing linguistic practices of circus trainers, equestrians, and dog obedience enthusiasts. All that philosophically suspect language is necessary to keep the humans alert to the fact that somebody is at home in the animals they work with.

Just *who* is at home must permanently be in question. The recognition that one cannot *know* the other or the self, but must ask in respect for all of time who and what are emerging in relationship is the key. That is true for all true lovers, of whatever species. Theologians describe the power of the "negative way of knowing" God. Because Who/What Is is infinite; a finite being, without idolatry, can only specify what is not, i.e., not the projection of one's own self. Another name for that kind of "negative" knowing is *love*. I believe those theological considerations are powerful for knowing dogs, especially for entering into a relationship, like training, worthy of the name of love.

I believe that all ethical relating, within or between species, is knit from the silk-strong thread of ongoing alertness to otherness-in-relation. We are not one, and being depends on getting on together. The obligation is to ask who are present and who are emergent. We know from recent research that dogs, even kennel-raised puppies, do much better than generally more brilliant wolves or human-like chimpanzees in responding to human visual, indexical (pointing), and tapping cues in a food-finding test. Dogs' survival in species and individual time regularly depends on their reading humans well. Would that we were as sure that most humans respond at better than chance levels to what dogs tell them. In fruitful contradiction, Hearne thinks that the intention-ascribing idioms of experienced dog handlers can prevent the kind of literalist anthropomorphism that sees furry humans in animal bodies and measures their worth in scales of similarity to the rights-bearing, humanist subjects of Western philosophy and political theory.

Her resistance to literalist anthropomorphism and her commitment to significant otherness-in-connection fuel Hearne's arguments against animal rights discourse. Put another way, she is in love with the cross-species achievement made possible by the hierarchical discipline of companion animal training. Hearne finds excellence in action to be beautiful, hard, specific, and personal. She is against the abstract scales of comparison of mental functions or consciousness that rank organisms in a modernist great chain of being and assign privileges or guardianship accordingly. She is after specificity.

The outrageous equating of the killing of the Jews in Nazi Germany, the Holocaust, with the butcheries of the animal-industrial complex, made famous by the character Elizabeth Costello in J. M. Coetzee's novel *The Lives of Animals,* or the equating of the practices of human slavery with the domestication of animals makes no sense in Hearne's framework. Atrocities, as well as precious achievements, deserve their own potent languages and ethical responses, including the assignment of priority in practice. Situated emergence of more livable worlds depends on that differential sensibility. Hearne is in love with the beauty of the ontological choreography when dogs and the humans converse with skill, face to face. She is convinced that this is the choreography of "animal happiness," a title of another of her books.

In her famous blast in *Harper's* magazine in September 1991 titled "Horses, Hounds and Jeffersonian Happiness: What's Wrong with Animal Rights?" Hearne asked what companion "animal happiness" might be. Her answer: the capacity for satisfaction that comes from striving, from work, from fulfillment of possibility. That sort of happiness comes from bringing out what is within, i.e., from what Hearne says animal trainers call "talent." Much companion animal talent can only come to fruition in the *relational* work of training. Following Aristotle, Hearne argues that this happiness is fundamentally about an ethics committed to "getting it right," to the satisfaction of achievement. A dog and handler discover happiness together in the labor of training. That is an example of emergent naturecultures.

This kind of happiness is about yearning for excellence and having the chance to try to reach it in terms recognizable to concrete beings, not to categorical abstractions. Not all animals are

*The Companion Species Manifesto*

alike; their specificity—of kind and of individual—matters. The specificity of their happiness matters, and that is something that has to be brought to emergence. Hearne's translation of Aristotelian and Jeffersonian happiness is about human–animal flourishing as conjoined mortal beings. If conventional humanism is dead in postcyborg and postcolonial worlds, Jeffersonian caninism might still deserve a hearing.

Bringing Thomas Jefferson into the kennel, Hearne believes that the origin of rights is in committed relationship, not in separate and preexisting category identities. Therefore, in training, dogs obtain "rights" in specific humans. In relationship, dogs and humans construct "rights" in each other, such as the right to demand respect, attention, and response. Hearne described the sport of dog obedience as a place to increase the dog's power to claim rights against the human. Learning to obey one's dog honestly is the daunting task of the owner. Her language remaining relentlessly political and philosophical, Hearne asserts that in educating her dogs she "enfranchises" a relationship. The question turns out not to be, What are animal rights, as if they existed preformed to be uncovered but, How may a human enter into a rights relationship with an animal? Such rights, rooted in reciprocal possession, turn out to be hard to dissolve; and the demands they make are life changing for all the partners.

Hearne's arguments about companion animal happiness, reciprocal possession, and the right to the pursuit of happiness are a far cry from the ascription of "slavery" to the state of all domestic animals, including "pets." Rather, for her the face-to-face relationships of companion species make something new and elegant possible; and that new thing is not human guardianship in place of ownership, even as it is also not property relations as conventionally understood. Hearne sees not only the humans but also the dogs as beings with a species-specific capacity for moral understanding and serious achievement. Possession—property—is about reciprocity and rights of access. If I have a dog, my dog has a human; what that means concretely is at stake. Hearne remodels Jefferson's ideas of property and happiness even as she brings them into the worlds of tracking, hunting, obedience, and household manners.

Hearne's ideal of animal happiness and rights is also a far cry from the relief of suffering as the core human obligation to animals. Human obligation to companion animals is much more exacting than that, even as daunting as ongoing cruelty and indifference are in this domain too. The ethic of flourishing described by the environmental feminist Chris Cuomo is close to Hearne's approach. Something important comes into the world in the relational practice of training; all the participants are remodeled by it. Hearne loved language about language; she would have recognized metaplasm all the way down.

## Apprenticed to Agility

Dear Vicki Hearne,

Watching my Aussi-mix dog Roland with you lurking inside my head last week made me remember that such things are multidimensional and situational, and describing a dog's temperament takes more precision than I achieved. We go to an off-leash, cliff-enclosed beach almost every day. There are two main classes of dogs there: retrievers and metaretrievers. Roland is a metaretriever. Roland will play ball with Rusten and me once in a while (or anytime we couple the sport with a liver cookie or two), but his heart's not in it. The activity is not really self-rewarding to him, and his lack of style shows it. But metaretrieving is another matter entirely. The retrievers watch whoever is about to throw a ball or stick as if their lives depended on the next few seconds. The metaretrievers watch the retrievers with an exquisite sensitivity to directional cues and microsecond of spring. These metadogs do not watch the ball or the human; they watch the ruminant-surrogates-in-dog's-clothing. Roland in meta-mode looks like an Aussie–Border Collie mock-up for a lesson in Platonism. His forequarters are lowered, forelegs slightly apart with one in front of the other in hair-trigger balance, his hackles in midrise, his eyes focused, his whole body ready to spring into hard, directed action. When the re-

## The Companion Species Manifesto

trievers sail out after the projectile, the metaretrievers move out of their intense eye and stalk into heading, heeling, bunching, and cutting their charges with joy and skill. The good metaretrievers can even handle more than one retriever at a time. The good retrievers can dodge the metas and still make their catch in an eye-amazing leap—or surge into the waves, if things have gone to sea.

Since we have no ducks or other surrogate sheep or cattle on the beach, the retrievers have to do duty for the metas. Some retriever people take exception to this multitasking of their dogs (I can hardly blame them), so those of us with metas try to distract our dogs once in a while with some game they inevitably find much less satisfying. I drew a mental Gary Larson cartoon on Thursday watching Roland, an ancient and arthritic Old English Sheepdog, a lovely red tricolor Aussie, and a Border Collie mix of some kind form an intense ring around a shepherd-lab mix, a plethora of motley Goldens, and a game pointer who hovered around a human, who—liberal individualist in Amerika to the end—was trying to throw his stick to his dog only.

*Correspondence with Gail Frazier,*
*agility teacher, May 6, 2001*

Hi Gail,

Your pupils, Roland Dog and I, got 2 Qualifying scores in Standard Novice this weekend at the USDAA trial!

*The Companion Species Manifesto*

Our early-morning Gamblers game on Saturday was a bad bet. And we were a disgrace to Agilitude in our Jumpers run, which finally happened at 6:30 p.m. Saturday evening. In our defense, after getting up at 4 a.m. on three hours sleep to get to Hayward for the trial, we were lucky to be standing by then, much less running and jumping. Both Roland and I ran totally separate jumpers courses, neither being the one the judge had prescribed. But our Standard runs Saturday and Sunday were both real pretty, and one earned us a 1st place ribbon. Roland's feet and my shoulders seemed born to dance together.

Cayenne and I head for Haute Dawgs in Dixon next Saturday for her first fun match. Wish us luck. There are so many ways to crash and burn on a course, but so far all of them have been fun, or at least instructive. Dissecting our respective runs Sunday afternoon in Hayward, one man and I were laughing at the cosmic arrogance of U.S. culture (in this case, ourselves), in which we generally believe both that mistakes have causes and that we can know them. The gods are laughing.

### The Game Story

Partly inspired by horse jumping events, the sport of dog agility first appeared at the Crufts dog show in London in February 1978 as entertainment during the break after the obedience championship and before the group judging. Also in agility's pedigree was police dog training, which began in London in 1946 and used obstacles like the high, inclined A-frame that the Army had already adopted for its canine corps. Dog Working Trials, a demanding British competition that included three-foot-high bar jumps, six-foot-high panel jumps, and nine-foot broad jumps, added a third strand in agility's parentage. For early agility games, teeter-totters were scavenged from children's playgrounds and coal mine ventilation shafts were put into service as tunnels. Men—many "guys who worked down the coal mines and wanted a bit of fun with their dogs," in the words of U.K. dog trainer and agility historian John Rogerson in Brenda Fender's series on "History of Agility" in *Clean Run Magazine*—were the original enthusiasts for these activities. Crufts and television, sponsored by Pedigree Pet Foods, assured that human gender and class would be as variable in the sport as the lineage of its equipment.

Immensely popular in Britain, agility spread around the world even faster than dogs had dispersed globally after their domestication. The United States Dog Agility Association (USDAA) was founded in 1986. By 2000, agility attracted thousands of addicted participants in hundreds of meets around the country. A weekend event typically draws three hundred or more dogs and handlers, and many teams trial more than once a month and train at least weekly. Agility flourishes in Europe, Canada, Latin America, Australia, and Japan. Brazil won the Fedération Cynologique Internationale's World Cup in 2002.

*The Companion Species Manifesto*

The USDAA's Grand Prix event is televised; its videotapes are devoured by agility enthusiasts for new moves by the great dog–handler teams and new course layouts devised by devious judges. Weeklong training camps attended by hundreds of students working with famous handler-instructors occur in several states.

Evidenced in *Clean Run,* the sport's glossy monthly magazine, agility is becoming ever more technically demanding. A course is made up of twenty or so obstacles like jumps, six-foot-high A-frames, twelve weave poles in series, teeter-totters, and tunnels arranged in patterns by judges. Games (called things like Snooker, Gamblers, Pairs, Jumpers with Weaves, Tunnelers, and Standard) involve different obstacle configurations and rules and require diverse strategies. Players see the courses for the first time the day of the event and walk through them for ten minutes or so to plan their runs. Dogs have not seen the course until they are actually running it. Humans give signals with voice and body; dogs navigate the obstacles at speed in the designated order. Scores depend on time and accuracy. A run typically takes a minute or less, and events are decided by fractions of seconds. Agility relies on fast-twitch muscles, skeletal and neural! Depending on the sponsoring organization, a dog–human team runs from two to eight events in a day. Recognition of obstacle patterns, knowledge of moves, skill on hard obstacles, and perfection of coordination and communication between dog and handler are keys to good runs.

*Cayenne Pepper leaps through the tire obstacle. Courtesy of Tien Tran Photography.*

Agility can be expensive; travel, camping, entry fees, and training easily run to $2,500 a year. To be good, teams need to practice several times a week and to be physically fit. The time commitment is not trivial for dogs or people. In the United States, middle-aged, middle-class, white women dominate the sport numerically; the best players internationally are more various in gender, color, and age, but probably not class. All sorts of dogs play and win, but particular breeds—Border Collies, Shetland Sheepdogs, Jack Russell Terriers—excel in their jump-height classes. The sport is strictly amateur, staffed and played

*The Companion Species Manifesto*

by volunteers and participants. Ann Leffler and Dair Gillespie, sociologists in Utah who study (and play) the sport, talk about agility in terms of "passionate avocations" that problematize the interface between public/private and work/leisure. I work to convince my sports writer father that agility should nudge football aside and take its rightful place on television with world-class tennis. Beyond the simple, personal fact of joy in time and work with my dogs, why do I care? Indeed, in a world full of so many urgent ecological and political crises, *how* can I care?

Love, commitment, and yearning for skill with another are not zero-sum games. Acts of love like training in Vicki Hearne's sense breed acts of love like caring about and for other concatenated, emergent worlds. That is the core of my companion species manifesto. I experience agility as a particular good in itself and also as a way to become more worldly, i.e., more alert to the demands of significant otherness at all the scales that making more livable worlds demands. The devil here, as elsewhere, is in the details. Linkages are in the details. Someday I will write a big book called, if not *Birth of the Kennel* in honor of Foucault, then *Notes of a Sports Writer's Daughter* in honor of another of my progenitors, to argue for the myriad strands connecting dogs to the many worlds we need to make flourish. Here, I can only suggest. To do that, I will work tropically by appealing to three phrases that Gail Frazier, my agility teacher, regularly uses with her students: "You left your dog"; "Your dog doesn't trust you"; and "Trust your dog."

## *The Companion Species Manifesto*

*Roland sails over a bar jump. Courtesy of Tien Tran Photography.*

These three phrases return us to Marco's story, Garrett's positive bondage, and Hearne's harsh beauty. A good agility teacher, like mine, can show her students exactly where they left their dogs and exactly what gestures, actions, and attitudes block trust. It's all quite literal. At first, the moves seem small, insignificant; the timing too demanding, too hard; the consistency too strict, the teacher too demanding. Then, dog and human figure out, if only for a minute, how to get on together, how to move with sheer joy and skill over a hard course, how to communicate, how to be honest. The goal is the oxymoron of disciplined spontaneity. Both dog and handler have to be able to take the initiative and to respond obediently to the other. The task is to become coherent enough in an incoherent world to engage in a joint dance of being that breeds respect and response in the flesh, in the run, on the course. And then to remember how to live like that at every scale, with all the partners.

## V. BREED STORIES

So far this manifesto has foregrounded two sorts of time–space scales co-constituted by human, animal, and inanimate agencies: (1) evolutionary time at the level of the planet Earth and its naturalcultural species, and (2) face-to-face time at the scale of mortal bodies and individual lifetimes. Evolutionary stories attempted to calm my political people's fears of biological reductionism and, with my colleague in science studies, Bruno Latour, interest them in the much more lively ventures of naturecultures. Love and training stories tried to honor the world in its irreducible, personal detail. At every repetition, my manifesto works fractally, reinscribing similar shapes of attention, listening, and respect.

It is time to sound tones on another scale, namely, historical time on the scale of decades, centuries, populations, regions, and nations. Here, I borrow from Katie King's work on femin-

*The Companion Species Manifesto*

ism and writing technologies, where she asks how to recognize emergent forms of consciousness, including methods of analysis, implicated in globalization processes. She writes about distributed agencies, "layers of locals and globals," and political futures yet to be actualized. Dog people need to learn how to inherit difficult histories in order to shape more vital multispecies futures. Attention to layered and distributed complexity helps me to avoid both pessimistic determinism and romantic idealism. Dogland turns out to be built from layers of locals and globals.

I need feminist anthropologist Anna Tsing to think about scale-making in dogland. She interrogates what gets to count as the "global" in transnational financial wheeling and dealing in contemporary Indonesia. She sees not preexisting entities already in the shapes and scales of frontiers, centers, locals, or globals, but instead "scale-making" of world-making kinds, in which reopening what seemed closed remains possible.

Finally, I translate—literally, move over to dogland—Neferti Tadiar's understanding of experience as living historical labor, through which subjects can be structurally situated in systems of power without reducing them to raw material for the Big Actors like Capitalism and Imperialism. She might forgive me for including dogs among those subjects, and she would give me the human-dog dyad at least provisionally. Let us see if telling histories of two divergent kinds of dogs—livestock guardian dogs (LGDs) and herders—and of institutionalized breeds emergent

*The Companion Species Manifesto*

from those kinds—Great Pyrenees and Australian Shepherds—as well as of dogs of no fixed breed or kind, can help shape a potent worldly consciousness in solidarity with my feminist, antiracist, queer, and socialist comrades, that is, with the imagined community that can only be known through the negative way of naming, like all the ultimate hopes.

In that negative way, I tell declarative stories trippingly. There are myriad origin and behavior stories about breeds and kinds of dogs, but not all narratives are born equal. My mentors in dogland taught me their breed histories, which I think honor both lay and scientific documentary, oral, experimental, and experiential evidence. The following stories are composites that, interpellating me into their structures, show something important about companion species living in naturecultures.

### Great Pyrenees

Guardian dogs associated with sheep- and goat-herding peoples go back thousands of years and cover wide swaths of Africa, Europe, and Asia. Local and long-range migrations of millions of grazers, shepherds, and dogs to and from markets and to and from winter and summer pastures—from the Atlas Mountains of North Africa, crossing Portugal and Spain, throughout the Pyrenean mountains, across southern Europe, over into Turkey, into Eastern Europe, across Eurasia, and through Tibet and into

China's Gobi Desert—have literally carved deep tracks into soil and rock. In their rich book *Dogs*, Raymond and Lorna Coppinger compare these tracks to the carving of glaciers. Regional livestock guardian dogs developed into distinct kinds in both appearance and attitude, but sexual communication always linked adjacent or traveling populations. The dogs that developed in higher, more northern, colder climates are bigger than those that took shape in Mediterranean or desert ecologies. The Spanish, English, and other Europeans brought their big mastiff-type and little shepherd-type dogs to the Americas in that massive gene exchange known as the conquest. Such interconnecting but far from randomly mixed populations are ecological and genetic population biologists' dreams or nightmares, depending on that hard thing called history.

Post mid-nineteenth-century kennel club breeds of LGDs with closed stud books derive from varying numbers of individuals collected from regional kinds, such as the Pyrenean Mastiff in the Basque area of Spain, the Great Pyrenees in Basque regions of France and Spain, the Maremma in Italy, the Kuvasz in Hungary, and the Anatolian Sheepdog in Turkey. The controversies about the genetic health and functional significance of these closed "island" populations called breeds rage in dogland. A breed club is partly analogous to a managing association for endangered species, for which population bottlenecks and disruption of past genetic natural and artificial selection systems require sustained, organized action.

*The Companion Species Manifesto*

157

Traditionally, LGDs protect flocks from bears, wolves, thieves, and strange dogs. LGDs often work with herding dogs in the same flocks, but the canines' jobs are different and their interactions limited. Regionally distinct, smallish herding dogs were everywhere, including hoards of collie types we will hear more about when I turn to Australian Shepherds. Peasant-shepherds across the huge land mass and time span of herding economies applied strong functional standards to their dogs that directly affected survival and breeding opportunities and shaped type. Ecological conditions also shaped the dogs and sheep independently of human intentions. Meanwhile, the dogs, employing different criteria, surely exercised their own sexual proclivities with their neighbors when they had the chance.

Guardian dogs do not herd sheep; they protect them from predators, mainly by patrolling boundaries and energetically barking to warn off strangers. They will attack and even kill intruders who insist, but their ability to calibrate their aggression to the level of the threat is legendary. They also perfect a repertoire of distinct barks for kinds and levels of alerts. Livestock guardian dogs tend to have low prey drive; and little of their puppy play involves chase, gather, head, heel, and grab/bite games. If they start to play like that with livestock or each other, the shepherd dissuades them. Those not dissuaded don't stay in the LGD gene pool. Working LGDs show the ropes to youngsters; lacking that, a knowledgeable human must help a lone puppy or older dog learn to be a good guardian—or, conversely, ignorantly set the neophyte up for failure.

Livestock guardian dogs tend to make lousy retrievers, and their biosocial predilections and upbringing conspire to deafen most to the siren songs of higher obedience competition. But they are capable of impressive independent decision-making in a complex historical ecology. Stories about LGDs' helping ewes give birth and licking the newborn lamb clean dramatize the dogs' capacity to bond with their charges. A livestock guardian dog, like a Great Pyrenees, might pass the day lounging among the sheep and the night patrolling, happily alert for trouble.

LGDs and herders tend to learn things with differential ease or difficulty. Neither kind of dog can really be taught to do their core jobs, much less the other dog's work. Dogs' functional behavior and attitudes can and must be directed and encouraged—trained, in that sense—but a dog with little joy in chasing and gathering and no deep interest in working with a human cannot be shown how to herd skillfully. Herders have strong prey drive from puppyhood. Choreographed with human herders and their herbivores, controlled components of that predation pattern, minus the kill and dissect parts, are precisely what herding is. Similarly, a dog with little passion for territory, anemic suspicion of intruders, and dim pleasure in social bonding cannot be shown from scratch how to think well about these things, even with the world's biggest clicker.

Guarding flocks in Europe since at least Roman times, large

*The Companion Species Manifesto*

white guardian dogs appear in French records over the centuries. In 1885–86, Pyrenean Mountain Dogs were registered with the Kennel Club in London. In 1909, the first Pyrs were brought to England for breeding. In his monumental 1897 encyclopedia *Les races des chiens,* Conte Henri de Bylandt dedicated several pages to describing Pyrenean guardian dogs. Forming rival clubs at Lourdes and Argeles, in 1907 two groups of French fanciers bought mountain dogs that they regarded as worthy and "purebred." Complete with the romantic idealization of peasant-shepherds and their animals characteristic of capitalist modernization and class formations that make such life ways nearly impossible, discourses of pure blood and nobility haunt modern breeds like the undead.

World War I destroyed both French clubs and most of the dogs. Working guardian dogs in the mountains were ravaged by war and depression, but they had already lost most of their jobs by the turn of the nineteenth century due to the extirpation of bears and wolves. Pyrs had become more likely to hang out as village dogs and be sold to tourists and collectors than put to work guarding flocks. In 1927, the diplomat, show judge, breeder, and native of the Pyrenees, Bernard Senac-Lagrange joined the few remaining fanciers to found the Réunion des Amateurs de Chiens Pyreneans and write the description that remains the foundation for current standards.

In the 1930s, serious collecting by two wealthy women, Mary Crane from Massachusetts (Basquaerie Kennels) and

Mme. Jeanne Harper Trois Fontaine from England (De Fontenay Kennel), brought many dogs out of France. The American Kennel Club recognized Great Pyrenees in 1933. World War II took another toll on the remaining LGDs in the Pyrenees and wiped out most of the French and Northern European registered kennel dogs. Asking how closely related they were and which left offspring, Pyr historians have tried to figure out how many dogs Mary Crane, Mme. Harper, and a few others bought, both from villagers and from fanciers. As few as thirty dogs, many related to each other, contributed in any continuing way to the gene pool of Pyrs in the United States. By the end of World War II, the only sizable Pyr populations in the world were in the United Kingdom and the United States, although the breed later recovered in France and northern Europe, with some exchange between U.S. and European breeders. The continuing existence of the dogs was largely due to the passionate show enthusiasts and breeders of the dog fancy. From 1931, when Mary Crane started collecting until the 1970s, very few U.S. Pyrs worked as livestock guardian dogs.

That changed with emerging approaches to predator control in the western United States in the early 1970s. Loose dogs killed lots of sheep. Coyotes also killed livestock; and they were ferociously poisoned, trapped, and shot by ranchers. Catherine de la Cruz—who got her first Pyr show bitch, Belle, in 1967 and was mentored in Great Pyrenees by Ruth Rhoades, the "mother superior" in the breed in California, who also taught Linda

Weisser—lived on a dairy ranch in Sonoma County. This middle-class, West Coast Pyr scene marks important differences in the breed's culture and future.

In 1972, a University of California–Davis scientist called de la Cruz's mother to talk about predator losses. The agribusiness research university and the U.S. Department of Agriculture were beginning to take nontoxic methods of predator control seriously. Environmental and animal rights activists were making their voices heard in public consciousness and national policy, including federal restrictions on using poisons to kill predators. De la Cruz's Belle hung out with the dairy cows between dog shows; that ranch never had any trouble with predators. De la Cruz relates that "the light went on in her head." The Great Pyrenees Standard describes the dogs guarding flocks from bears and wolves, although that was more the symbolic narrative of show fanciers than description of what any of them had seen. Whatever else it also does, the written standard in an institutionalized breed is about ideal type and origin narrative. In her own origin story, de la Cruz tells that she began to think that the Pyrs she knew might be able to guard sheep and cows from dogs and coyotes.

De la Cruz gave some puppies to northern California sheep people she knew. From there, she and a few other Pyr breeders, including Weisser, placed dogs (including some adults) on ranches and tried to figure out how to help the dogs become effective Predator Control Dogs, as they were called then. The

Mary Crane (left) in July 1967 at the Great Pyrenees Club of America National Specialty Show in Santa Barbara, California. The dog next to Mrs. Crane is Armand (Ch. Los Pyrtos Armand of Pyr Oaks), who won the stud dog class that day. Next to him are his two daughters: Impy, who went Reserve Winners bitch, and Drifty, who was Best of Opposite Sex. Linda Weisser is with Drifty, who died without offspring. Weisser's "dog of my heart," Impy has descendants in almost all U.S. West Coast kennels. Through a son, Armand is behind Catherine de la Cruz's working ranch stock. Courtesy of Linda Weisser and Catherine de la Cruz.

dairy farm was converted to sheep ranching, and de la Cruz became part of the woolgrowers' association. In the late 1970s, she met Margaret Hoffman, a woman active in the woolgrowers' group who wanted dogs to repel coyotes. Hoffman got Sno-Bear from de la Cruz, bred more dogs, and placed 100 percent of them in working homes. In an interview with me in November 2002, de la Cruz talks about "making every possible mistake," experimenting with socializing and caring for working dogs, staying in close touch with the ranchers, and cooperating with UC Davis and Department of Agriculture people in research and placement.

In the 1980s, Linda Weisser and Evelyn Stuart, part of the Great Pyrenees Club of America committee to revise the standard, made sure that the functional, working dogs were prominently in view. By the 1980s, de la Cruz, still showing dogs in conformation, was placing working Pyrs around the country. A few of the dogs came in from the pastures, got their baths, won championships, and went right back to work. The "dual-purpose dog" became a moral and practical ideal in Pyr breeding and breed education. Mentoring to achieve this ideal involves all kinds of labor—and labor-intensive—practices, including managing high-quality Internet listservs like the Livestock Guardian Dog Discussion List and the stockguard topic section of the Great Pyrenees Discussion List. Lay expertise, volunteer labor, and collaborating communities of practice are crucial. Not least, every working Pyr in the United States comes through a pet and show home history of more than four decades. Companion species and emergent naturecultures appear everywhere I look.

Beginning in the mid-1970s, first Jeffrey Green and then also Roger Woodruff of the U.S. Sheep Experiment Station of the U.S. Department of Agriculture (USDA) in Dubois, Idaho, are key actors in this story. Their first guardian dog was a Komondor (Hungary), and they then worked with Akbash (Turkey) and

Pyrs. My Pyr informants discuss these men with tremendous respect. Urging ranchers to try out the guardian dogs, the USDA men solicited breeders' help and treated them as colleagues. For example, Woodruff and Green gave a special seminar on LGDs at the Great Pyrenees Club of America National Specialty show in Sacramento in 1984. Another piece of the story of the re-emergence of working LGDs in North America is Hal Black's early-1980s study of Navajo sheep herding practices with their effective mongrel dogs to glean lessons for other ranchers.

Rancher reeducation was a big part of the USDA project, and Pyr people engaged that process energetically. Steeped in the modernization ideologies of the science-based, land grant universities and agribusiness, ranchers tended to see dogs as old-fashioned and commercial poisons as progressive and profitable. Dogs are not a quick fix; they require changed labor practices and investments of time and money. Working with ranchers to effect change has been modestly successful.

In 1987 and 1988, the USDA project bought about a hundred guardian dog puppies from around the United States, most of them Pyrs. The USDA scientists agreed to the breed club people's insistence on spaying and neutering the dogs placed through the project, which kept at least those dogs out of puppy mill production and other breeding practices that the club people believe harmful to the dogs' well-being and genetic health. To reduce the risk of hip dysplasia in the working dogs, all of the parents of the pups had their hips checked by X-rays. By the late

*The Companion Species Manifesto*

1980s, surveys indicated that more than 80 percent of ranchers found their guardian dogs—especially their Great Pyrenees—to be an economic asset. By 2002, a few thousand LGDs are in charge of the protection of sheep, llamas, cattle, goats, and ostriches throughout the United States.

Raymond and Lorna Coppinger and their associates at Hampshire College's New England Farm Center, beginning with Anatolian Shepherds brought from Turkey in the late 1970s, also did research and placed hundreds of LGDs on American farms and ranches. Raymond Coppinger has a PhD in the tradition of Niko Tinbergen's ethology legacy at Oxford University, and the Coppingers also have a serious history in racing sled dogs. The Coppingers have always been more in the public eye and better known by scientists, other than those directly involved in LGD work, than the lay breeders whom I emphasize in my story. The Coppingers dissent on many points from the view of guardian dogs held by my Pyr people. The Hampshire College project did not sterilize dogs they placed. Believing that the social environment during maturation was the only crucial variable in shaping an effective stock guardian, they did not generally take breed distinctions seriously. The Hampshire project placed younger puppies, taught a different view of biosocial development and genetic behavioral predilections, and handled the mentoring of people and dogs differently.

Most Pyr people did not cooperate with the Coppingers, and animosity dates from the start. Effectively, the Coppingers had little access to Great Pyrenees, where the breed club ethic was strong. I cannot evaluate the differences here, and the reader can find the Coppingers' views in *Dogs*. In that book, there is no mention of the Pyr people, including that they were placing livestock guardian dogs and cooperating with Jeff Green and Roger Woodruff from the start. Readers will also not learn, as they could in a 1990 USDA publication, that in a 1986 survey of four hundred people, involving 763 dogs, conducted by the University of Idaho, Great Pyrenees made up 57 percent of the population. Pyrs and Komondors, another breed whose people did not contribute to the Hampshire project, accounted for 75 percent of the working LGDs in the study. That study and others show that Pyrs tend to get the highest marks of any breed for job success. That includes biting fewer people and injuring fewer livestock. In a study of yearling dogs involving fifty-nine Pyrs and twenty-six Anatolian Shepherds, 83 percent of Pyrs got a score of "good" compared to 26 percent of the Anatolians.

The introduction, from blasted peasant-shepherd econo-mies, of Basque Pyrenean Mountain Dogs, who were nurtured in the purebred dog fancy, onto the ranches of the U.S. West to protect Anglo ranchers' xenobiological cattle and sheep on the grasslands habitat (where few native grasses survive) of buffalo once hunted by Plains Indians riding Spanish horses—along with the study of contemporary reservation Navajo sheep-herding cultures deriving from Spanish conquest and mission-ization—ought to offer enough historical irony for any compan-ion species manifesto. But there is more. Two efforts to bring back extirpated predator species rehabilitated from the status of

*The Companion Species Manifesto*

*Great Pyrenees puppy learning the job among the sheep. Courtesy of Linda Weisser and Catherine de la Cruz.*

vermin to natural wildlife and tourist attraction, one in the Pyrenean mountains and one in the national parks of the American West, will lead us further into the web.

The Endangered Species Act in the United States gives the Department of the Interior jurisdiction over reintroduction of the gray wolf to parts of its previous range, such as Yellowstone National Park, where fourteen Canadian wolves were released in 1995 in the midst of the country's largest elk and buffalo populations. Migrating Canadian wolves began showing up in Montana on their own initiative. In 1995–96, fifty-two more wolves were released in Idaho and Wyoming. About seven hundred wolves live in the northern Rocky Mountains in 2002. By and large, ranchers remain unreconciled, even though they get full monetary compensation for stock losses and stock-killing wolves are removed or killed by the Fish and Wildlife Service of the Department of the Interior. According to Jim Robbins's *New York Times* report on December 17, 2002 (page D3), 20 percent of the closely managed wolves wear electronic monitoring collars. Coyote numbers are down; wolves kill them. Elk numbers are down. That makes hunters unhappy but pleases ecologists worried about damage from herbivores deprived of their predators. Tourists—and the businesses that serve them—are very happy. More than a hundred thousand tourist wolf sightings have been logged on car safaris in the Lamar Valley in Wyoming. No tourists have been killed, but national figures in 2002 showed that two hundred cattle, five hundred sheep, seven llamas, one horse, and forty-three dogs have been. Who were those forty-three dogs?

Some of them were ill-prepared Great Pyrenees. The Department of the Interior put wolves in Yellowstone National Park against ranchers' wishes; without coordination with the Department of Agriculture LGD people in Idaho; and without, I suspect, even imagining talking to knowledgeable Pyr breeders, who are also late-middle-aged white women who show their gorgeous dogs in conformation. Interior and Agriculture are worlds apart in technoscientific culture. The wolves spilled out of park boundaries. Wolves, livestock, and dogs all got killed, maybe needlessly. Wildlife officials have killed more than 125 errant wolves; ranchers have illegally shot at least dozens more. Wildlife conservationists, tourists, ranchers, bureaucrats, and communities got polarized, maybe needlessly. Better companion species relations needed to be formed all around, from the start, among the humans and the nonhumans.

Dogs are social and territorial; wolves are social and territorial. Experienced LGDs in large enough established groups might be able to deter northern gray wolves from munching on livestock. But bringing Pyrs to the scene after the wolves have set up shop or using too few and inexperienced dogs are sure recipes for disaster for both canid species and for weaving together wildlife and ranching ethics. The group Defenders of Wildlife has bought Pyrs for ranchers experiencing losses to wolves; the wolves seem actively drawn to and kill the dogs as intruding competitors on wolf real estate. Practices that might have led wolves to respect organized dogs were not in place; it might be too late for LGDs to be effective actors in wolf flourishing and rancher–conservationist alliances. Maybe the wolves will control the coyotes while the Pyrs are protected indoors at night.

Meanwhile, restoration ecology has its European flavors. In the Pyrenees, the French government has introduced European Brown Bears from Slovakia, where the postcommunist tourist industry makes a tidy sum promoting bear watching, to fill the empty niche left by killing the previous ursine residents. French Pyr fanciers, such as the goat farmer Benoit Cockenpot of du Pic de Viscos kennel, work to get the dogs back in the mountains telling the Slovakian bears the proper postmodern order of things. The French Pyr fanciers are learning about working LGDs from their U.S. colleagues. The French government offers farmers a free guardian dog. But insurance reimburses farmers for animals lost to predators, and that is turning out to be more attractive than daily taking care of dogs. Guardian dogs have a harder time competing with the insurance apparatus than repelling bears.

Away from multispecies conservation and farm politics, Pyrs never stopped excelling as show dogs and pets. However, the breed's numerical expansion as both workers and pets has meant considerable escape from the breed club's control, much less the control of a viable peasant-shepherd economy, into the hells and limbos of commercial puppy production and backyard breeding. Indifference to health; ignorance of behavior, socialization, and training; and cruel conditions are all too frequent. Within the breed clubs, controversy reigns over what constitutes responsible breeding, especially when the hard-to-digest topics of genetic diversity and population genetics in purebred dogs are on the menu. Overuse of popular sires, secrecy about dogs' problems, and lusting for show ring championships at the expense of other values are practices known to imperil dogs. Too many people still do it. Love of dogs forbids it, and I have met many of these lovers in my research. These are the people who get dirty and knowledgeable in all the worlds where their dogs live — on farms, in labs, at shows, in homes, and wherever else. I want their love to flourish; that is one reason I write.

## *The Companion Species Manifesto*

## Australian Shepherds

The herding breed known in the United States as the Australian Shepherd, or Aussies, raises just as many complexities as Great Pyrenees; I will sketch only a few. My point is simple: knowing and living with these dogs means inheriting all of the conditions of their possibility, all of what makes relating with these beings actual, all of the prehensions that constitute companion species. To be in love means to be worldly, to be in connection with significant otherness and signifying others, on many scales, in layers of locals and globals, in ramifying webs. I want to know how to live with the histories I am coming to know.

If anything is certain about Australian Shepherd origins, it is that no one knows how the name came about, and no one knows all of the kinds of dogs tied in the ancestry of these talented herders. Perhaps the surest thing is that the dogs should be called the United States Western Ranch Dog. Not "American," but "United States." Let me explain why that matters, especially since most (but far from all) of the ancestors are probably varieties of collie types that emigrated with their people from the British Isles to the East Coast of North America from early colonial times on. The California Gold Rush and the aftermath of the Civil War are the keys to my regional national story. These epic events made the American West into part of the United States. I don't want to inherit these violent histories, as Cayenne, Roland, and I run our agility courses and conduct our oral affairs; that's why I have to tell them. Companion species cannot afford evolutionary, personal, or historical amnesia. Amnesia will corrupt sign and flesh and make love petty. If I tell the story of the Gold Rush and the Civil War, then maybe I can remember the other stories about the dogs and their people—stories about immigration, indigenous worlds, work, hope, love, play, and the possibility of cohabitation through reconsidering sovereignty and ecological developmental naturecultures.

Romantic origin stories about Aussies have late-nineteenth- and early-twentieth-century Basque herders bringing their little blue merle dogs with them in steerage as they headed, via sojourn in Australia herding merino sheep from Spain, for the ranches of California and Nevada to tend the sheep of a timeless pastoral West. "In steerage" gives the game away; working-class men in steerage were in no position to bring their dogs, to Australia or to California. Besides, the Basques who immigrated to Australia did not become herders but sugar cane workers; and they did not go Down Under until the twentieth century. Not necessarily shepherds before, the Basques came to California, sometimes via South America and Mexico, in the nineteenth century with the millions lusting for gold and ended up herding sheep to feed other disappointed miners. The Basques also established great restaurants, heavy on lamb dishes, in Nevada on what became the interstate highway system after World War II. The Basques got their sheep dogs from among local working herding dogs, who were a mixed lot, to say the least.

*The Companion Species Manifesto*

Spanish missions favored sheep ranching to "civilize" Native Americans, but in her online version of Aussie history, Linda Rorem notes that by the 1840s the number of sheep (not to mention Native people) in the Far West had greatly declined. Discovery of gold radically and permanently changed the food economy, politics, and ecology of the region. Large sheep flocks were transported by sailing them from the East Coast around the Horn, driving them overland from the Midwest and New Mexico, and shipping them from that "nearby" white settler colony with a colonial pastoral economy, Australia. Many of these sheep were merinos, originally of Spanish origin, but coming to Australia through Germany, as a gift from Spain's king to Saxony, which developed a thriving colonial export trade in sheep.

What the Gold Rush began the aftermath of the Civil War finished, with its vast influx of Anglo (and some African American) settlers to the West and the military destruction and containment of Native Americans and consolidations of expropriated land from Mexicans, Californios, and Indians.

All of these movements of sheep also meant movements of their herding dogs. These were not the guardian dogs of the old Eurasian pastoral economies, with their established market routes, seasonal pasturages, and local bears and wolves—which were, nonetheless, heavily depleted. The settler colonies in Australia and the United States adopted an even more aggressive attitude to natural predators—building fences around most of Queensland to keep out dingoes and trapping, poisoning, and shooting anything with serious canine teeth that moved upon the land in the U.S. West. Guardian dogs did not appear in the U.S. western sheep economy until after these tactics became illegal in the queer times of effective environmental movements.

The herding dogs accompanying the immigrant sheep from both the East Coast and Australia were mainly of the old working collie/shepherd types. These were strong, multipurpose dogs with a "loose eye" and upstanding working posture—rather than with a sheep trial-selected, Border Collie hard eye and crouch—from which several kennel club breeds derive. Among the dogs coming to the U.S. West from Australia were the frequently merle-colored "German Coulies," who look a lot like modern Australian Shepherds. These were British-derived, all-purpose herding "collies," called "German" because German settlers lived in an area of Australia where these dogs were common. Dogs that look like contemporary Aussies might have gotten their name early from being associated with flocks arriving on boats from Down Under, whether or not they came on those ships. Or, associated with later immigrant dogs, these types might have started being called "Australian Shepherds" as late as World War I. Written records are scarce. And there wasn't a "purebred" in sight for a long time.

There were, however, identifiable lines in California, Washington, Oregon, Colorado, and Arizona developing by the 1940s that became registered Australian Shepherds, beginning in

*The Companion Species Manifesto*

1956. Registration was not common until the mid-to-late-1970s. The range of types was still wide, and styles of dogs were associated with particular families and ranches. Curiously, a rodeo performer from Idaho named Jay Sisler is part of the story of molding a kind of dog into a contemporary breed, complete with its clubs and politics. Over twenty years, Sisler's "blue dogs" were a popular rodeo trick show. He knew the parents of most of these dogs, but that is as deep as genealogy got in the beginning. Sisler got his dogs from various ranchers, several of whose Aussies became foundation stock of the breed. Among the identified 1,371 dogs out of 2,046 ancestors in her ten-generation pedigree, I count seven Sisler dogs in my Cayenne's family. (Many with names like "Redding Ranch Dog" and "Blue Dog," 6,170 out of more than a million ancestors are known in her twenty-generation tree; that leaves a few gaps.)

An amazing trainer of the type Vicki Hearne would have loved, Sisler considered Keno, whom he got around 1945, to be his first really good dog. Keno contributed offspring to what became the breed; but the Sisler dog who made the biggest impact (percentage ancestry) to the current population of Aussies was John, a dog with unknown antecedents who wandered one day onto the Sisler ranch and into written pedigrees. There are many such stories of foundation dogs. They could all be microcosms for thinking about companion species and the invention of tradition in the flesh, as well as the text.

The Aussie parent club, the Australian Shepherd Club of

America (ASCA), was founded in Tucson by a small group of enthusiasts in 1957. ASCA wrote a preliminary standard in 1961 and a firm one in 1977 and got its own breed club registry going in 1971. Organized in 1969, the ASCA Stock Dog Committee organized herding trials and titles; and working ranch dogs began their considerable reeducation for the trial ring. Conformation competitions and other events became popular, and sizable numbers of Aussie people saw American Kennel Club (AKC) affiliation as the next step. Other Aussie people saw AKC recognition as the road to perdition for any working breed. The pro-AKC people broke away to found their own club, the United States Australian Shepherd Association (USASA), which got full AKC recognition in 1993.

All of the biosocial apparatus of modern breeds emerged—including savvy lay health and genetics activists, scientists researching illnesses common in the breed and perhaps establishing companies to market resultant vet biomedical products, Aussie-themed small businesses, performers passionate about the dogs in agility and obedience, both suburban weekend and rural ranching stock dog trialers, search and rescue workers, therapy dogs and their people, breeders committed to maintaining the versatile dog they inherited, other breeders enamored of big-coated show dogs with untested herding talent, and much more. C. A. Sharp, with her kitchen-table-produced *Double Helix Network News* and the Australian Shepherd Health & Genetics Institute that she helped found—not to mention her

*Beret's Dogon Grit winning High in Sheep at the 2002 Australian Shepherd Club of America National Stock Dog Finals, Bakersfield, California. Courtesy of Gayle Oxford, Glo Photo.*

reflection on her own practices as a breeder and her adoption of a too-small Aussie rescue pooch after the death of the last dog of her breeding—embodies for me the practice of love of a breed in its historical complexity.

Cayenne's breeders, Gayle and Shannon Oxford in California's Central Valley, are active in both the USASA and ASCA. Committed to breeding and training working stock dogs and also showing in conformation and agility, the Oxfords taught me about "the versatile Aussie," which I see as analogous to the Pyr people's "dual-purpose" or "whole dog" discourse. These idioms work to prevent the splitting up of breeds into ever more isolated gene pools, each dedicated to a specialists' limited goal, whether that be agility sports, beauty, or something else. The bedrock test of an Australian Shepherd, however, remains the ability to herd with consummate skill. If "versatility" does not start there, the working breed will not survive.

## A Category of One's Own

Anyone who has done historical research knows that the undocumented often have more to say about how the world is put together than do the well pedigreed. What do contemporary companion species relations between humans and "unregistered" dogs in technoculture tell us about both inheriting—or perhaps better, inhabiting—histories and also forging new possibilities? These are the dogs who need "A Category of One's Own," in honor of Virginia Woolf. Author of the famous feminist tract *A Room of One's Own*, Woolf understood what happens when the impure stroll over the lawns of the properly registered. She also understood what happens when these marked (and marking) beings get credentials and an income.

Generic scandals get my attention, especially the ones that ooze racialized sex and sexualized race for all the species involved. What should I call the categorically unfixed dogs, even if I stay only in America? Mutts, mongrels, All-Americans, random-bred dogs, Heinz 57s, mixed breeds, or just plain dogs? And why should categories for dogs in America be in English? Not just "the Americas," but also the United States is a highly polyglot world. Above, concentrating on Great Pyrenees and Australian Shepherds, I had to suggest the conundrums of inheriting local and global histories in modern breeds by a couple of shaggy-dog stories. Similarly, here I cannot begin to plumb the histories of all the sorts of dogs that fit into neither functional kind nor institutionalized breed. And so, I will offer only one story, but one that ramifies further into webs of worldly complexity at each retelling. I will tell about Satos.

*Sato* is slang in Puerto Rico for a street dog. I learned this fact in two places: on the Internet at www.saveasato.org and in Twig Mowatt's moving essay in the Fall 2002 issue of the glossy dog cultures magazine, *Bark*. Both of these sites landed me squarely in the naturecultures of what gets politely called "modernization." *Sato* is just about the only Spanish word I learned in either site; that cued me into the direction of the semiotic and material traffic in this zone of dogland. I also figured out that Satos are capitalized, in lexical convention and monetary investment, in the process of moving from the hard streets of the southern "developing world" to the "forever homes" of the enlightened north.

At least as important, I learned that I am interpellated into this story in mind and heart. I cannot disown it by calling attention to its racially tinged, sexually infused, class-saturated, and colonial tones and structures. Again and again in my manifesto, I and my people need to learn to inhabit histories, not disown them, least of all through the cheap tricks of puritanical critique. In the Sato story, there are two kinds of superficially opposed temptations to puritanical critique. The first is to indulge in the colonialist sentimentality that sees only philanthropic (philocanidic?) rescue of the abused in the traffic of dogs from Puerto Rican streets to no-kill animal shelters in the United States and from there to proper homes. The second is to indulge in historical structural analysis in a way that denies both emotional bonds and material complexity and so avoids the always messy participation in action that might improve lives across many kinds of difference.

About ten thousand Puerto Rican dogs have made the transition from street life to suburban homes since 1996 when airline worker Chantal Robles of San Juan teamed up with Karen Fehrenbach, visiting the island from Arkansas, to set up the Save-a-Sato Foundation. The facts that led them to action are searing. Millions of fertile and usually diseased and starving dogs scavenge for a meal and shelter in Puerto Rico's impoverished neighborhoods, construction sites, garbage dumps, gas stations, fast food parking lots, and drug sale zones. The dogs are rural and urban, big and little, recognizably from an institutionalized breed and plainly of no breed at all. They are mostly young—feral dogs don't tend to get very old; and there are lots of puppies, both abandoned by people and born to street bitches. Official animal shelters in Puerto Rico mainly kill the dogs and cats surrendered to them or collected in their sweeps. Sometimes these swept-up animals are owned and cared for; but they live rough, vulnerable to complaint and official action. Conditions in the municipal shelters are the stuff of an animal rights horror show.

Very many dogs of all sorts in Puerto Rico are, of course, well cared for. The poor as well as the wealthy cherish animals. But if people abandon a dog, they are far more likely to let the pooch loose than bring him or her to an underfunded and poorly staffed "shelter" that is certain to kill its charges. Furthermore, the class-, nation-, and culture-based animal welfare ethic of sterilizing dogs and cats is not widespread in Puerto Rico (or in much of Europe and many places in the United States). Mandatory sterilization and reproductive control have a very checkered history in Puerto Rico, even when one restricts one's historical memory to policies for nonhuman species. At the very least, the notion that the only proper dog is a sterile dog—except for those in the care of responsible (in whose view?) breeders— brings us smashing into the world of biopower and its technocultural apparatus in the metropole and the colonies. Puerto Rico is both metropole and colony.

None of this removes the fact that fertile feral dogs have sex, whelp lots of puppies they can't feed, and die of awful diseases in great pain and large numbers. It's not just a narrative. To make matters worse, Puerto Rico is no more free than the United

States of damaged, abusing people of all social classes who inflict dire mental and physical injuries on animals both deliberately and indifferently. Homeless animals, like homeless people, are fair game in the free trade—or maybe better, free fire—zones.

The action taken by Robles, Fehrenbach, and their supporters is, to me, as inspiring as it is disturbing. They established and run a private shelter in San Juan that functions as a halfway house for dogs on their way to mostly international adoption. (But Puerto Rico is part of the United States, or is it?) The demand in Puerto Rico for these dogs is slight; that is not a natural fact, but a biopolitical one. Anyone who has thought about human international adoption knows that. The Save-a-Sato Foundation raises money, trains volunteers to bring dogs (and some cats) to the shelter without further traumatizing them, organizes Puerto Rican veterinarians who treat and sterilize animals for free, socializes the future adoptees in manners proper to the north, prepares papers for them, and arranges with the airlines to ship about thirty dogs per week on commercial flights to a network of no-kill shelters in several states, mostly in the Northeast. Post-9/11, tourists flying out of San Juan are recruited to claim crates of emigrating dogs as their personal baggage so that the antiterrorist apparatus does not shut down the rescue pipeline.

The foundation runs an English-language website to inform its potential adopting audience and to link support groups to people who take the dogs into, in the idiom of the website, their "forever families." The website is full of successful adoption accounts, preadoption horror stories, before and after photos, invitations to take action and to contribute money, information for finding a Sato to adopt, and useful links to dogland cyber-culture.

A person in Puerto Rico can become a member of the Save-a-Sato Foundation by rescuing a minimum of five dogs per month. Volunteers mainly pay whatever it costs out of their own pockets. They find, feed, and gentle dogs before urging them into crates and taking them to the halfway house. Puppies and youngsters are the first priority, but not the only ones picked up. Dogs who are too sick to get well are euthanized, but many severely injured and ill dogs recover and get placed. All sorts of people become volunteers. The website tells about one elderly woman on Social Security living close to homelessness herself who recruited homeless people to gentle and collect dogs, for whom she paid five dollars each out of her meager funds. Knowing the genre of such a story does not mute its power—or its truth. The photos on the site seem to be mostly of middle-class Puerto Rican women, but heterogeneity in the Save-a-Sato Foundation is not reserved for the dogs.

The airplane is an instrument in a series of subject-transforming technologies. The dogs who come out of the belly of the plane are subject to a different social contract from the one they were born into. However, not just any Puerto Rican stray is

likely to get its second birth from this aluminum womb. Smallish dogs, like girls in the human scene, are the gold standard in
the dog adoption market. U.S. fear of aggression from the Other
knows few bounds, and certainly not those of species or sex. To
follow this point, we need to get from the airport to the excellent
shelter in Sterling, Massachusetts, which has placed more than
two thousand Satos (and about a hundred cats) since it joined
the program in 1999. Once again, I find my bearings in dogland's
exuberant cyberculture (www.sterlingshelter.org).

Animal shelters in the U.S. Northeast in general have too few
dogs in the ten-to-thirty-five-pound range to fill the demand.
Being the owner (or guardian) of a midsized, sterilized, rescue-
derived, well-behaved dog in the United States confers high
status in much of dogland. Some of this status comes from pride
in not succumbing to the eugenic discourses that continue to
luxuriate in purebred dog worlds. But adoption of a street or
thrown-away dog, mutt or not, hardly removes one from the
swamps of class- and culture-rooted "improving" ideologies,
familial biopolitics, and pedagogical fashions. Indeed, eugenics
and the other improving discourses of "modern" life have so
many shared ancestors (and living siblings) that the coefficient
of inbreeding exceeds that of even father–daughter couplings.

Adopting a shelter dog takes a lot of work, a fair amount of
money (but not as much as it costs to prepare the dogs), and a
willingness to submit to a governing apparatus sufficient to ac-
tivate the allergies of any Foucauldian or garden-variety liber-

tarian. I support that apparatus—and many other kinds of institutionalized power—to protect classes of subjects, including dogs. I also vigorously support adopting rescue and shelter animals. And so my dyspepsia at recognizing where all this comes from will have to be endured rather than relieved.

Good shelters get lots of requests for Sato dogs. Getting such a dog keeps people from buying from pet stores and supporting the puppy mill industry. The Sterling shelter tells us that 99 percent of puppies brought to it from the United States are medium to large dogs, all of whom get adopted. Many largish puppies and youngsters come into the Sterling haven from the Homebound Hounds Program, which imports thrown-away dogs to the Northeast from cooperating shelters in the U.S. South—another area of the world where the ethic of sterilizing dogs and cats is not secure, to say the least. Still, people looking for smaller shelter dogs are largely out of luck in the domestic market. These folks' family enlargement strategies require different layers of locals and globals. However, just as with international adoption of children, it is not easy to get an imported dog. Detailed interviews and forms, home visits, references from friends and veterinarians, promises to educate the dog properly, counseling from on-site trainers, proof of home ownership or written documentation from landlords that pets are allowed, and then long waiting lists: all this and more is normal. The goal is a permanent home for the dogs.

The means is a kinship-making apparatus that reaches into and draws from the history of "the family" in every imaginable way, literally. Proof of the effectiveness of the companion-species, family-making apparatus is to be found in a little narrative analysis. Adoption success stories regularly refer to siblings and other multispecies kin as mom, dad, sister, brother, aunt, uncle, cousin, godfather, etc. Purebred adoption stories do the same thing, and these adoption/ownership processes involve many of the same documentary and social instruments before one can qualify to get a dog. It is nearly impossible—and generally irrelevant—to read from the stories what species is being referred to. A pet bird is the sister of a new dog, and the human baby brother and aged cat aunt all are represented to relate to the human adults of the house as moms and/or dads. Heterosexuality is not germane; heterospecificity is.

I resist being called the "mom" to my dogs because I fear infantilization of the adult canines and misidentification of the important fact that I wanted dogs, not babies. My multispecies family is not about surrogacy and substitutes; we are trying to live other tropes, other metaplasms. We need other nouns and pronouns for the kin genres of companion species, just as we did (and still do) for the spectrum of genders. Except in a party invitation or a philosophical discussion, *significant other* won't do for human sexual partners; and the term performs little better to house the daily meanings of cobbled-together kin relations in dogland.

But perhaps I worry about words too much. I have to admit that it is not clear that the conventional kin idioms in use in U.S. dogland refer to age, species, or biological reproductive status

*The Companion Species Manifesto*

much at all (except to require that most of the nonhumans be sterile). Genes are not the point, and that surely is a relief. The point is companion species–making. It's all in the family, for better and for worse, until death do us part. This is a family made up in the belly of the monster of inherited histories that have to be inhabited to be transformed. I always knew that if I turned up pregnant, I wanted the being in my womb to be a member of another species; maybe that turns out to be the general condition. It's not just mutts, in or out of the traffic of international adoption, who seek a category of one's own in significant otherness.

I yearn for much more reflection in dogland about what it means to inherit the multispecies, relentlessly complex legacy that crosses evolutionary, personal, and historical time scales of companion species. Every registered breed, indeed every dog, is immersed in practices and stories that can and should tie dog people into myriad histories of living labor, class formations, gender and sexual elaborations, racial categories, and other layers of locals and globals. Most dogs on Earth are not members of institutionalized breeds. Village dogs and rural and urban feral dogs carry their own signifying otherness for the people they live among, and not just for people like me. Nor are mutts or so-called "random-bred" dogs in the "developed world" like the functional kinds of dogs that emerged in economies and ecologies that no longer flourish. Puerto Rican strays called *Satos* become members of Massachusetts "forever families" out of histories of stunning complexity and consequence. In current naturecultures, breeds might be a necessary, if deeply flawed, means to continue the useful kinds of dogs they came from. Current U.S. ranchers have more to fear from real estate developers from San Francisco or Denver than from wolves, no matter how far they get from the parks, or from Native Americans, no matter how effective they are in court.

In my own personal-historical natureculture, I know in my flesh that the largely middle-class white people of Pyr and Aussie land have an as yet unarticulated responsibility to participate in reimagining grasslands ecologies and ways of life that were blasted in significant part by the very ranching practices that required the work of these dogs. Through their dogs, people like me are tied to indigenous sovereignty rights, ranching economic and ecological survival, radical reform of the meat industrial complex, racial justice, the consequences of war and migration, and the institutions of technoculture. It's about, in Helen Verran's words, "getting on together." When "purebred" Cayenne, "mixed-breed" Roland, and I touch, we embody in the flesh the connections of the dogs and the people who made us possible. When I stroke my landmate Susan Caudill's sensuous Great Pyrenees, Willem, I also touch relocated Canadian gray wolves, upscale Slovakian bears, and international restoration ecology, as well as dog shows and multinational pastoral economies. Along with the whole dog, we need the whole legacy, which is, after all, what makes the whole com-

*Willem and Cayenne play in spring 2000. Photograph by the author.*

panion species possible. Not so oddly, all those wholes are non-Euclidean knots of partial connections. Inhabiting that legacy without the pose of innocence, we might hope for the creative grace of play.

*From "Notes of a Sports
Writer's Daughter," June 2000*

Ms. Cayenne Pepper has shown her true species being at last. She's a female Klingon in heat. You may not watch much television or be a fan of the *Star Trek* universe like I am, but I'll bet the news that Klingon females are formidable sexual beings, whose tastes run to the ferocious, has reached everyone in the federated planets. The Pyr on our land, the intact twenty-month-old Willem, has been Cayenne's playmate since they were both puppies, beginning at about four months of age. Cayenne was spayed when she was six and a half months old. She's always happily humped her way down Willem's soft and inviting backside, starting at his head with her nose pointed to his tail, while he lies on the ground trying to chew her leg or lick a rapidly passing genital area. But during our Memorial weekend stay on the Healdsburg land, things heated up, put mildly. Willem is a randy, gentle, utterly inexperienced, adolescent male soul. Cayenne does not have an estrus hormone in her body (but let us not forget those very much present adrenal cortices pumping out so-called androgens that get lots of the credit for juicing up mammalian desire in males and females). She is, however, one turned-on little bitch with Willem, and he is INTERESTED. She does not do this with any other dog, "intact" or not. None of their sexual play has anything to do with remotely functional heterosexual mating behavior—no efforts of Willem to mount, no presenting of an attractive female backside, not much genital sniffing, no whining and pacing, none of all that reproductive stuff. No, here we have pure polymorphous perversity that is so dear to the hearts of all of us who came of age in the 1960s reading Norman O. Brown.

The 110-pound Willem lies down with a bright look in his eye. Cayenne, weighing in at thirty-five pounds, looks positively crazed as she straddles her genital area on top of his head, her nose pointed toward his tail, presses down and wags her backside vigorously. I mean hard and fast. He tries for all he's worth to get his tongue on her genitals, which inevitably dislodges her from the top of his head. It looks a bit like the rodeo, with her riding a bronco and staying on as long as possible. They have slightly different goals in this game, but both are committed to the activity. Sure looks like eros to me. Definitely not agape. They keep this up for about three minutes to the exclusion of any other activity. Then they go back to it for another round. And another. Susan's and my laughing, whether raucous or discrete, does not merit their attention. Cayenne growls like a female Klingon during the activity, teeth bared. Remember how many times the half-Klingon B'Elanna Torres on *Star Trek Voyager* put her human lover Tom Paris in sickbay? Cayenne's playing, but oh my, what a game. Willem is earnestly intent. He is not a Klingon, but what feminists of my generation would call a considerate lover.

Their youth and vitality make a mockery of reproductive heterosexual hegemony, as well as of abstinence-promoting gonadectomies. Now, I, of all people, who have written infamous books about how we Western humans project our social orders and desires onto animals without scruple, should know better than to see confirmation of Norman O. Brown's *Love's Body* in my spayed Aussie dynamo and Susan's talented Landscape Guardian Dog with that big, sloppy, velvety tongue. Still, what else could be going on? Hint: this is not a game of fetch or chase.

No, this is ontological choreography, which is that vital sort of play that the participants invent out of the histories of body and mind they inherit and that they rework into the fleshly verbs that make them who they are. They invented this game; this game remodels them. Metaplasm, once again. It always comes back to the biological flavor of the important words. The word is made flesh in mortal naturecultures.

## BIBLIOGRAPHY

Ackerley, J. R. 1956. *My Dog Tulip*. Great Britain: Secker and Warburg.

Althusser, Louis. 1970. *Lenin and Philosophy, and Other Essays*. Trans. Ben Brewster. New York: Monthly Review Press.

Clark, Mary T. ed. 2000. *An Aquinas Reader: Selections from the Writings of Thomas Aquinas*. New York: Fordham University Press.

Australian Shepherd Club of America. 1978. *Yearbook 1957–77*. Los Angeles: Australian Shepherd Club of America.

——. 1985. *Yearbook 1978–82*. Los Angeles: Australian Shepherd Club of America.

Black, Hal. 1981. "Navajo Sheep and Goat Guarding Dogs: A New World Solution to the Coyote Problem." *Rangelands* 3 (6): 235–38.

Brown, Norman O. 1966. *Love's Body*. New York: Vintage.

Budiansky, Stephen. 1992. *The Covenant of the Wild: Why Animals Chose Domestication*. New York: William Morrow.

*The Companion Species Manifesto*

Butler, Judith. 1990. *Gender Trouble: Feminism and the Subversion of Identity and Bodies*. New York: Routledge.

Coetzee, J. M. 2001. *The Lives of Animals*. Princeton, N.J.: Princeton University Press.

Coppinger, Raymond, and Lorna Coppinger. 2001. *Dogs: A Startling New Understanding of Canine Origin, Behavior, and Evolution*. New York: Scribner.

Cuomo, Chris. 1998. *Feminism and Ecological Communities*. New York: Routledge.

Darwin, Charles, Paul Ekman, and Phillip Prodger. 1998. *The Expression of the Emotions in Man and Animals*. 3rd ed. London: Harper Collins. Originally published in 1872.

de Bylandt, Conte Henri. 1897. *Les races de chiens, leurs origines, points, descriptions, types, qualités, aptitudes et défauts*. Bruxelles: Vanbuggenhoudt frères. Reprinted 2013.

de la Cruz, Catherine. Interview with author, Santa Rosa, Calif., November 2002.

——. N.d. "GPRNC Profiles: Catherine de la Cruz." http://www.sonic.net/~cdlcruz/Rescue/RD/BoardProfiles/catherine.htm. Accessed August 2015.

Fender, Brenda. 2004. "History of Agility." *Clean Run Magazine* (July): 32–36; (August): 28–33; (September): 26–29. http://www.cleanrun.com/index.cfm/category/702/history-of-agility.htm. Accessed August 2015.

Foucault, Michel. 1973. *Birth of the Clinic*. Trans. Alan Sheridan. London: Routledge.

Freedman, Adam, et al. 2014. "Genome Sequencing Highlights the Dynamic Early History of Dogs." *PLoS Genetics* 10 (8): e1004631.

*The Companion Species Manifesto*

Garrett, Susan. 2002. *Ruff Love*. South Hadley, Mass.: Clean Run Productions.

Gilbert, Scott F., and David Epel. 2015. *Ecological Developmental Biology*. 2nd ed. Sunderland, Mass.: Sinauer.

Gillespie, Dair, Ann Leffler, and Elinor Lerner. 2001. "If It Weren't for My Hobby, I'd Have a Life: Dog Sports, Serious Leisure, and Boundary Negotiations." Paper presented at the American Sociological Association section on Animals and Society, Anaheim, Calif.

Goldsworthy, Andy, and David Craig. 1999. *Arch*. New York: Abrams.

Great Pyrenees Library. http://www.greatpyreneeslibrary.com/. Accessed August 2015.

Green, Jeffrey, and Robert Woodruff. 1999. "Livestock Guarding Dogs: Protecting Sheep from Predators." U.S. Department of Agriculture, Agriculture Information Bulletin, no. 588.

Haraway, Donna. 1985. "Manifesto for Cyborgs: Science, Technology, and Socialist Feminism in the 1980s." *Socialist Review* 80: 65–108.

———. 2008. *When Species Meet*. Minneapolis: University of Minnesota Press.

Hearne, Vicki. 1982. *Adam's Task*. New York: Random House.

———. 1994. *Animal Happiness*. New York: Harper Collins.

———. 1991. "Horses, Hounds and Jeffersonian Happiness: What's Wrong with Animal Rights?" *Harper's* (September): 59–64. http://harpers.org/archive/1991/09/whats-wrong-with-animal-rights/. Accessed August 2015. Available online with a new prologue at www.dogtrainingarts.com.

King, Katie. 1994. "Feminism and Writing Technologies." *Configurations* 2 (1): 89–106.

Koehler, William R. 1996. *The Koehler Method of Dog Training*. New York: Howell Book House.

## The Companion Species Manifesto

Latour, Bruno. 1993. *We Have Never Been Modern*. Cambridge, Mass.: Harvard University Press.

———. 2004. "Why Has Critique Run Out of Steam? From Matters of Fact to Matters of Concern." *Critical Inquiry* 30 (2): 225–48.

Margulis, Lynn. 1991. "Symbiogenesis and Symbionticism." In *Symbiosis as a Source of Evolutionary Innovation: Speciation and Morphogenesis,* ed. L. Margulis and R. Fester, 1–14. Boston: MIT Press.

———, and Dorian Sagan. 2002. *Acquiring Genomes: A Theory of the Origin of Species*. New York: Basic Books.

McCaig, Donald. 1984. *Nop's Trials*. New York: Lyons Press.

———. 1994. *Nop's Hope*. New York: Lyons Press.

McFall-Ngai, Margaret. 2014. "Divining the Essence of Symbiosis: Insights from the Squid-Vibrio Model." *PLoS Biology* 12 (2): e1001783.

Morey, Darcy. 2010. *Dogs: Domestication and the Development of a Social Bond*. Cambridge, U.K.: Cambridge University Press.

Mowatt, Twig. 2002. "Second Chance Satos." *Bark* 20 (Fall).

Noske, Barbara. 1989. *Humans and Other Animals: Beyond the Boundary of Anthropology*. London: Pluto Press.

Podberscek, Anthony L., Elizabeth S. Paul, and James A. Serpell, eds. 2000. *Companion Animals and Us*. Cambridge, U.K.: Cambridge University Press.

Princehouse, Patricia. N.d. "History of the Pyrenean Shepherd." http://pyrshep1.homestead.com/pshistory.html. Accessed August 2015.

Robbins, Jim. 2002. "More Wolves and New Questions, in Rockies." *New York Times,* December 17, D3. http://www.nytimes.com/2002/12/17/science/more-wolves-and-new-questions-in-rockies.html. Accessed August 2015.

Rorem, Linda. 1987. "A View of Australian Shepherd History." Stockdog Library. Originally published in *Dog World*. Revised 2007, 2010.

## *The Companion Species Manifesto*

http://www.workingaussiesource.com/stockdoglibrary/rorem_his
tory_article.htm. Accessed August 2015.

Russell, Edmund. 2004. "Introduction: The Garden in the Machine. Toward an Evolutionary History of Technology." In *Industrializing Organisms: Introducing Evolutionary History,* ed. Susan Schrepfer and
Philip Scranton, 1–18. London: Routledge.

Save-a-Sato Foundation. www.saveasato.org. Accessed August 2015.

Schwartz, Marion. 1997. *A History of Dogs in the Early Americas.* New
Haven: Yale University Press.

Scott, John Paul, and John L. Fuller. 1965. *Genetics and the Social Behavior
of the Dog.* Chicago: University of Chicago Press.

Serpell, James. 1986. *In the Company of Animals: A Study of Human-An-
imal Relationships.* Cambridge, U.K.: Cambridge University Press.

———, ed. 1995. *The Domestic Dog: Its Evolution, Behaviour, and Interactions with People.* Cambridge, U.K.: Cambridge University Press.

Sharp, C. A. 1993–2014. *Double Helix Network News.* Privately produced
four times/year, Fresno, Calif.

———. Australian Shepherd Health and Genetics Institute. http://www
.ashgi.org/home-page/about-ashgi/board-of-directors/ca-sharp.
Accessed August 2015.

Smuts, Barbara. 2000. "Encounters with Animal Minds." *Journal of Consciousness Studies* 8 (5–7): 293–309.

———. 2008. "Between Species: Science and Subjectivity." *Configurations*
14 (1–2): 115–26.

Strathern, Marilyn. 1991. *Partial Connections.* Lanham, Md.: Rowman
and Littlefield.

Tadiar, Neferti. 2009. *Things Fall Away: Philippine Historical Experience
and the Making of Globalization.* Durham, N.C.: Duke University
Press.

## *The Companion Species Manifesto*

Thompson, Charis. 2005. *Making Parents: The Ontological Choreography of Reproductive Technologies.* Cambridge, Mass.: MIT Press.

Tinbergen, Niko. 1953. *The Herring Gull's World.* London: Collins.

Tsing, Anna. 2012. "Unruly Edges: Mushrooms as Companion Species." *Environmental Humanities* 1: 141–54.

Verran, Helen. 2001. *Science and an African Logic.* Chicago: University of Chicago Press.

——. 2014. "Working with Those Who Think Otherwise." *Common Knowledge* 20 (3): 527–39.

Vilá, Carles, J. E. Maldonado, and R. K. Wayne. 1999. "Phylogenetic Relationships, Evolution, and Genetic Diversity of the Domestic Dog." *American Genetics Association* 90: 71–77.

Weisser, Linda. 2000. Interview with author, Olympia, Wash., December 29–30.

"Weisser, Linda, 1940–2011." *Great Pyrenees Club of America Bulletin* (2nd quarter): 12–13. http://gpcaonline.org/PDF/GPCA%20Q2%20 2011%20Bulletin.pdf. Accessed August 2015.

Whitehead, Alfred North. 1929. *Process and Reality.* New York: Macmillan.

Wilson, Cindy C., and Dennis Turner, eds. 1998. *Companion Animals in Human Health.* Thousand Oaks, Calif.: Sage.

Woolf, Virginia. 1929. *A Room of One's Own.* Oxford, U.K.: Oxford University Press.

*The Companion Species Manifesto*

# Companions in Conversation

The following conversation took place over a three-day period, May 11–13, 2014, at the home of Donna Haraway and Rusten Hogness in Santa Cruz, California. During the previous week, both participants had been involved in conferences referred to during the conversation—Cary Wolfe in the conference Sciences and Fiction, organized by the Center for the Study of the Novel at Stanford University, and Donna Haraway at the conference Anthropocene: Arts of Living on a Damaged Planet, organized jointly by Anna Tsing and colleagues in the anthropology department at the University of California at Santa Cruz and in the Aarhus University Research on the Anthropocene (AURA) network in Denmark.

## CYBORG BEGINNINGS

CARY WOLFE: I want to talk about the original context—in whatever way you would like (intellectual, institutional, political)—of the two manifestos, and how that shaped not just the composition and motivations behind the pieces, but also their reception, because obviously a lot of things have changed since 1983, but a lot of things have changed since the "Companion Species Manifesto," too.

I thought that might help us explore the afterlife of both of these pieces as they bear upon work that people are doing now. So let's start there.

DONNA HARAWAY: Let's start with the "Cyborg Manifesto." I was asked by the *Socialist Review* West Coast Collective, along with several other folks who had identified in various ways as socialist-feminist, Marxist-feminist—a fairly broad understanding of what those formations meant—in the early Reagan years, (we were in the early '80s then) to write a few pages envisioning what was possible, where to move, how to move now, in this conjuncture of what we subsequently look at as the Reagan–Thatcher years. You could no longer not know that the '60s were well and truly over, and that the great hopefulness of our politics and imaginations needed to come to terms with serious troubles within our own movements, within our larger historical moment. What did we think about it? The "Cyborg Manifesto" emerged partly from that invitation. Also, I was asked to prepare a paper, as delegate for the *Socialist Review,* to a meeting in (then) Yugoslavia of the New Left and post–New Left, of the Eastern European and Euro- and American parties. At that meeting I met extremely interesting other Marxist-feminists, as well as other folks attending and working at the conference. We experienced a kind of immediate bonding over issues of who was doing the Xeroxing and who was doing the speaking—those kinds of things, old-fashioned feminist issues that never go away.

The manifesto grew out of these multiple immediate contexts, but more, it also grew very much out of a sense of being a child of WWII, growing up with a brain educated by Sputnik—that is to say, the fact that the United States was in competition in the space race with the Soviet Union, that produced such things as the National Defense Education Act and textbook revision in the sciences, across the biological sciences and indeed in the social sciences. (My friend Susan Harding is writing about MACOS, Man a Course of Study, the fascinating middle-school curricular reform that came out of the same social conjunctures.)

I had just moved in the early '80s to Santa Cruz, from teaching at Johns Hopkins and before that at the University of Hawaii. The Applied Physics Laboratory at Hopkins and the Pacific Strategic Command in Hawaii made me see the military industrial complex as it is embodied, embedded, in elite research apparatuses and in real places. (There's much more to say about how teaching at Johns Hopkins shaped me, for example, learning the history of the School of Hygiene and Public Health.) I was personally shaped by the embedded institutional and political apparatus of these complex formations of capitalism, militarism, imperialism, and more. Baltimore was also where, with Nancy Hartsock, I experienced a vital Marxist-feminist collective, as well as the Baltimore Experimental High School, where my lover and later husband, Rusten Hogness, taught and where I came to read and embrace the anarcha-feminism of Marge Piercy. It was a very important period of time for me. I was teaching and learning the history of science, and that has remained very important to me.

And before that, Hawaii. I landed in Hawaii as a biology graduate student from Yale who was riveted by the way that biology is culture and practice, culture and politics, material natureculture—you know, phrases I formed later, but approaches that I was already deeply involved with. And then, in Honolulu, I was married to a gay man, Jaye Miller, who remained the friend of my heart all of his life. But obviously marriage was a bad idea—what were we doing? I decided we were engaged in a fairly innocent form of incest, we were sort of brother and sister—I don't know what we were doing *(laughs)*. There we were, from Yale and, you know, in Honolulu, you land on what looks like the New Haven Green, and you understand that this is the Yale–New Haven Green all over again, heir of those of Protestant formations, the Congregationalists, the missionaries, the sugar families, the commercial and religious and political apparatus of American Protestant hegemony. There we were: plantations, colonialism, racial formations, the Pacific Strategic Command in the middle of the Vietnam War, sexual and kin and gender experimentation, vital social movements, including the resurgence of Hawaiian sovereignty movements—all of it, and all of it materially built into the land.

The "Cyborg Manifesto" is a kind of coming together of understanding that I had been formed, as who I am in the world, out of these large and small, intimate and huge matters—way too big to comprehend but lived in the intimate tissues of your own friendships and politics and love affairs and so forth of post–WWII American hegemony. Lived particularly in the forms power took in information-saturated culture, information science–saturated culture and politics, in Command Control Communication Intelligence (C³I). C³I was central to the McNamara plan in the Vietnam War—the particular cybernetic rationalization of war, much of which was run from Hawaii, during the very period of indigenous Hawaiian sovereignty movements, struggles for feminism and reproductive and sexual freedom, and land and labor struggle movements, both Hawaiian and not, with the hotel industry following the movement of plantation agriculture out of Hawaii and the great expansion of the tourist industry. I was formed as a person out of all these things.

And I was and remained always profoundly in love with biology, the critters, the ways of knowing. All of this made me ever more aware of how the way we know the world, including ourselves, is situated historically in particular apparatuses for knowing, so that we know ourselves as a system—an information system, as a system divided by the division of labor. We know ourselves as a heat engine, we know ourselves as a telephone exchange. . . . These things are never mere metaphors—

*Companions in Conversation*

we really are historically crafted in these knowledge practices. These things may be made, but they are not made up. So the "Cyborg Manifesto" was an effort to come to terms with these converging, even imploding ways of understanding being in the world and being responsible in the world. I was writing as a feminist, a Marxist, a biologist, a teacher, a friend, whatever, at a certain historical moment.

cw: One of the fascinating things to me about the "Manifesto"—and I'm not exaggerating when I say this—is that I'm not sure I can think of any single document in my academic life that has been taken up more variously, let's just say *(laughs)*, by more different audiences (just staying within academia), for more different purposes, than the "Cyborg Manifesto." And in that way, it's a document with a different kind of life in many ways from the "Companion Species Manifesto." And that's also a product of when the piece was published and famously tracing—you're right, *at that moment*—those boundary breakdowns that you identified ...

DH: ... and recompositions.

cw: And recompositions. But I also think (and this sometimes falls out of some of the more, you might say, futuristic appropriations of the "Manifesto") it involves how you constantly circle back in the piece to embed all of this in the incredible transformation in the sciences at that time (as you put it, the understanding of biological entities in cybernetic terms as now being "not optional").

## Companions in Conversation

DH: That's right.

cw: But you also embed these plate-tectonic shifts in that discipline, and in cultural studies and feminism as well, within much longer stories, such as "the God-trick" that we're all familiar with.

DH: And in many ways, the sister paper to the "Cyborg Manifesto" is "Situated Knowledges."

cw: Right.

DH: But staying with the "Cyborg Manifesto," I didn't have the language then for saying these things this way, but critique was never enough, because love and rage are the affects, are my affectual relationship to being in the world in this time/space warp in which we find ourselves *now*—whatever you want to call it, this thick and fibrous *now*. How to truly love our age, and also how to somehow live and die well here, with each other? Also, the manifesto was shaped by the ongoing looping through a particular moment of women-of-color feminism, and the call to account by Chela Sandoval and others, of the overly white feminisms of many of "our" visions and understandings, many of *my* formations, of that period. The "Cyborg Manifesto" tries to live with and be accountable to racist formations in and out of feminism, accountable to the deep troubles of socialism in and out of formal Marxist analysis, and so forth. I needed somehow to stay with a nonsimplistic and always troubled sense of being within a politics and being for some worlds and not others.

*Companions in Conversation*

cw: Right. And yet I do think that one of the things that opened the "Cyborg Manifesto" to so many different audiences is a really, really important term in the piece, and that's the term *irony*.

DH: Yes.

cw: And that's also very early-'80s.

DH: Yes, non–self-identity . . .

cw: Right, that's also a very located term in the history of literary criticism of the period that we're talking about. But I think that the balance in the piece between all of these—between the heartfelt, deep, visceral commitments that you're voicing right now and being able to maintain this kind of ironic stance in relation to the figure of the cyborg, I think that . . .

DH: . . . that mattered.

cw: Well, I think what it did was to open the "Cyborg Manifesto" to a much broader audience doing many different kinds of appropriations that actually had nothing to do with feminism or Marxism or biology in the minds of the appropriators.

DH: Absolutely right. And, it opened it up to communities of practice, so that it's taken up by performance artists and many others. I had no idea. I mean, I certainly wasn't deliberately writing to those communities, but subsequently, because it was taken up, I've met people engaged in collaborative work of all sorts. And now I do write to and with those audiences. I probably wouldn't today call what I was doing *irony*, in part because the word has this complicated history. But every act of syntax is also a kind of fierce joke on our desire to clarify, to control, to know, to identify. But by the time you reach the end of a sentence, you've said at least six things that aren't true and you don't hold, but to get to the end of the sentence you don't have any choice. You can't simply say what you mean—that's not how language works.

cw: Right, and irony was shorthand for what you would develop as a much more thoroughgoing vocabulary involving lots of the figures that you use in the "Companion Species Manifesto."

DH: Right, absolutely. Well, and remember that when I wrote the "Cyborg Manifesto," I was a brand new faculty member in the History of Consciousness program at UC Santa Cruz, this very interesting formation. I was still trying to learn of lot of contemporary theory in the human sciences, mostly new to me, using words in sentences just to see if I could, like I was in grade school again. In many ways the "Cyborg Manifesto" was trying out some of the knowledges that hadn't been mine that I was getting from my colleagues and the graduate students in the program, and that came to be part of poststructuralism and deconstruction in various ways—some of the theories of Jakob von Uexküll and Roland Barthes and many others. That paper was also my coming to locate myself in my new playground.

cw: Right. In a set of new discourses . . .

DH: And I made a whole lot of mistakes that turned out sometimes to be kind of happy mistakes. Some of them I made on purpose, because I didn't want to use the stuff the way others seemed to be using it. And some of it was that I really didn't understand and made mistakes that ended up being interesting.

cw: Everybody, of course—especially given the history of your career—everybody thinks of the "Cyborg Manifesto" as a key document in the whole history of feminist thought. But less so in socialist thought. And that has to do, I think, less with what you wrote than with all the overdetermining forces of reception inside and outside the academy that really changed the fortunes of Marxism and socialism within the academy, which up to that point was still a very, very robust tradition. As we've talked about before, Fred Jameson was the first reader on my dissertation, and looking back I now see— and I've told this to many people—that in a way he is, ironically enough, the last European intellectual.

DH: It is ironic, isn't it?

cw: An intellectual of a certain tradition.

DH: Well, I read *The Political Unconscious* at about the time that I wrote the "Manifesto," too. Foucault was by then an old friend, but not yet Jameson.

cw: Yes, and so, to think about how the fortunes of these things we write depend upon these much larger and quite institutional forces, not just intellectual formations.

DH: Well, and you know, the East Coast *Socialist Review* collective hated the "Cyborg Manifesto" and the Berkeley-located Bay Area *Socialist Review* [SR] collective embraced it, largely because of Jeff Escoffier, who was a really lovely man, deeply political, and a great editor. The manifesto caused immediate controversy at *SR,* and it caused immediate controversy within feminisms of many kinds, not least because it adamantly refused an anti–science-and-technology stance or vocabulary. My cyborg would have none of that, but it also refused to be a blissed-out technobunny. It refused a nothing-but-critique approach to the vast things that the heavens know needed serious critique (and still do). The nothing-but-critique approach was a temptation in some crucial domains of feminism and New Left socialism. The "Cyborg Manifesto" was a deliberate in-your-face NO to that relation to science and technology, and that caused controversy from the get-go.

cw: Right, and that circles us back to the figure of irony in the piece, but it also accounts for the extremely long life of the "Cyborg Manifesto" in terms of its own relevance. Had you hewed to either of those narratives being held by the people that you upset, the "Cyborg Manifesto" . . .

DH: . . . would have had its moment.

## Companions in Conversation

cw: Exactly. Would have been a nice essay, and so on.

DH: But it remains disturbing, and it remains disturbing to me.

cw: Yes.


## WORKING TOWARD UNKNOWING

DH: Because it's actually a paper about "both/and," "yes/and," "no/but," "no/and," etc. It is a figure and a paper, a mode of working, and a statement of "Best I can tell, this is not just the way I work, this is how worlding works." And that both/and —but never in a kind of easy way; it's not additive—this kind of meeting each other across serious oppositional difference doesn't resolve into some kind of dialectical resolution. None of the cognitive technologies I inherited has ever been solace enough for that feeling that if you reach the period of the sentence, then you have moved into a precious place of "unknowing," through the relentless pressure of saying and feeling yes/but, both/and.

cw: That's another reason that the piece has had such longstanding relevance: when I go back and read the piece—and this connects you to characters like Foucault and others doing work in continental philosophy at that time, but also to different kinds of pragmatism in North America (mainly of the Left variety)—it was actually a rethinking of what politics is. What do we mean by "the political"? You could say, "What is political theory?" except part of the force of your point is that *political theory* isn't the term we want either, because we're really talking about these practices of constitution, as you would say in your later work. You have a much larger vocabulary for talking about getting on in the world with others, or staying with the trouble, and so forth.

DH: Yes, that we all keep thinking together.

cw: Yes, and so a huge achievement of the piece—and it's one that you can't just unilaterally make happen by being intentional about it (hence the importance of how the piece is written, its figural quality)—is that it created this echo chamber or seed bed (to mix my metaphors) for thinking about what "the political" is, which took another twenty or thirty years to fully get out into the world and hook up with other efforts that were going on elsewhere but were still constrained under labels like "socialism."

DH: Yes, no question. Or feminism, or antiracism.

cw: Or cybernetics, for that matter.

DH: Heaven knows . . .

cw: And so one of the achievements that would not have been possible without this kind of stance—as we said, *irony* is one word, but there are other ways of talking about it—is that it opened up a space in which those aspects of the essay could be taken up and developed by others.

*Companions in Conversation*

DH: And were.

CW: In a million different directions.

DH: I still get these emails from high school kids saying, "This was assigned to me, I don't understand it. Would you please explain it?" I mean, high school kids, oh my god! *(Both laugh.)* Mostly, I try to answer those emails, at least a little.

CW: Right—how long do we have here *(laughing)*?!

SITUATING COMPANION SPECIES

CW: I want to move on and talk for a bit about the "Companion Species Manifesto." Eventually I want to come back to both of the manifestos side by side to talk about how they've ramified in related but also very different ways into something that a lot of people are interested in right now—namely, biopolitical thought. But before we do that, I wanted to turn to the "Companion Species Manifesto" and ask you the same kind of question—and you talk about this explicitly in the piece—about the contexts of its composition, the motivations behind it, some of which I know were personal, some of which were political and institutional, but also, again, the context of its reception, because it's a very different moment in feminism, in academia more broadly, in cultural studies, and so on. There still wasn't what we now call "Cultural Animal Studies" or "Human Animal Studies."

DH: Or "Multispecies Studies."

CW: Or Multispecies Studies, which was well on the way to being composed and cobbled together, but . . .

DH: . . . it didn't quite exist yet.

CW: Yes, exactly.

DH: Well, like the "Cyborg Manifesto," the "Companion Species Manifesto" is situated in a historical conjuncture that is felt deeply personally and is simultaneously much more than personal. It is part of a *reworlding*—that science fiction term has been very important to me. It seems to me that it is a term necessary for ordinary thinking, way beyond whatever counts as science fiction, these reworldings. So the "Companion Species Manifesto" comes at a point of no longer being able to write or think without asking, Who are we here? What are we? Who and what are "we" that is not only human? What is it to be companion species at this historical conjuncture, and so what? Who lives and who dies, how, and so what? Here, in this conjuncture?

And *companion species* for me never meant just companion animals, although companion animals are among them, but, rather, the name was at least, like the cyborg, spin-outable, it could be spun out, like silk out of a spider's abdomen, multiple strong silk threads. We are companions, *cum panis*, at table together. We are those who are at risk to each other, who are each other's flesh, who eat and are eaten, and who get indigestion, who are, in Lynn Margulis's sense, in the symbiogenetic conjuncture of living and dying on Earth. We are in a systems world, as in the "Cyborg Manifesto," but more alert to *sympoietic* systems (not self-making, not autopoietic), making-in-symphony, making-with, never one, always looping with other worlds. And *species,* the relentlessly oxymoronic quality of a word that is both the ideal type, the coin, the *specie,* the money, the biological entity, the science fiction species, the detail that's a species *of* something else. *Species* is an inherently incredibly complex word; it just explodes with its incongruous multiple meanings.

CW: We could talk about the Norman O. Brown connection here *(laughs).*

DH: Yes, well, I mean, a multispecies *Love's Body,* heaven help us! *(Both laughing.)*

So, species is way more than my dog and me playing, *and,* simultaneously, it is me playing with my dog and being undone and redone by that. I found myself with this flaming talented youngster of another species, two weekends a month and several hours of training every week in addition, playing a game that neither of us invented, flaming through these sports fields in California in, of all places, the fairgrounds, with the NASCAR races and the railroad tracks, the quinceañera fiestas—the parties for the fifteen-year-old Latina girls—and in the fairgrounds, talk about being in the middle of California social, agricultural, industrial history! In the middle of the history of the expansion of the United States nation as it marches across conquered territory. I dare you to find a more potent place for being at risk to each other than the fairgrounds *(laughs).*

cw: In all senses of the word! *(Both laughing.)*

DH: Truly! Playing our game, me and my dog and my friends and their dogs, and trying to figure out who this "we" is that we become-with each other. Truly, who is this "we"? And it's simultaneously a moment when many of the ecofeminists and deep ecology people and animal rights people are making a claim on *us*. Composing a "we," too. They're composing, or proposing, a really important kind of question: Are we together here or not? I mean, what is the "here" and who are the "we," where critters are at stake to us and to each other? And they are at stake in the *Animal Industrial Complex,* which, remember, was Barbara Noske's term from the early '80s, around the period when the "Cyborg Manifesto" was composed. You know that because you were already beginning to work, had been working for years, at this intersection, or implosion—what's the right metaphor, after all?—of the questions of the flourishing of human and nonhuman critters in their entanglements.

cw: Yes, and at that point that was largely regarded, at least when I started, as nonserious work.

DH: Absolutely. Nobody took this stuff seriously, in universities anyway. Well, in fact, I gave a precursor of the "Companion

Species Manifesto" as a talk for the Cultural Studies Colloquium on my campus around 2002. A friend of mine, who has remained a close colleague and friend, came up and said, "I absolutely loved that talk, it was fantastic, but I hope you know that unlike the 'Cyborg,' this won't take off." Well *(they both laugh)*, I don't know if that turned out to be very prophetic. *(Still laughing.)* It's not like *companion species* takes off as such as a term, at this conjuncture, this worlding, or whatever it is we call it—"animal studies," "multispecies studies," "companion species studies." The question of the human/animal divide in all of it—well, the *multiple* divides, because they're not single divides—the question of the comings-together and the dividings of those who share (and bear) vitality and mortality, of those of us who are mortal creatures on this Earth in this historical moment: this has taken off in ways we could not imagine even a few years ago.

cw: Yes, and that sets a fundamentally different tone in the "Companion Species Manifesto" from the "Cyborg Manifesto." There's a sense of finitude, a sense of mortality. A palpable sense of the presence of life and death that's in a different register from what you get in the "Cyborg Manifesto."

dh: It's in a different register, and the tone of the writer is much more personal and vulnerable. The narrative voice in the "Companion Species Manifesto," the "I" in that work . . .

CW: Yes, a lot of people read the "Cyborg Manifesto" very much in the mode of performance, and that's very different from the voice you get later.

DH: It's a different voice. There are folks who asked, "Why did you drop your feminist, antiracist, and socialist critique in the 'Companion Species Manifesto'?" Well, it's not dropped. It's at least as acute, but it's produced very differently. There's a sense in which the "Companion Species Manifesto" grows more out of an act of love, and the "Cyborg Manifesto" grows more out of an act of rage.

CW: I don't think you drop the critique at all. In fact—and we'll talk about this in a minute—I think what happens is that what your friend called "socialist, antiracist, and feminist commitments" are sustained, but they're retooled within a context that I would call more thoroughgoingly biopolitical.

DH: I hope that's true.

CW: And that's a very different context from command-control-communication-intelligence and the military industrial complex, and the notes that you were sounding in the "Cyborg Manifesto." Not that those things go away, obviously, but . . .

DH: . . . they are configured differently.

CW: They ramify differently, I would say. They're made flesh differently, as you would put it in the "Companion Species Manifesto."

*Companions in Conversation*

One point of contact—and this is a deep subterranean connection between the two manifestos—is that a lot of people have noted, celebrated, how the manifesto begins with this kind of deep tongue kiss between you and Cayenne.

DH: That soft-porn moment. *(They both laugh.)*

CW: Famous or infamous—call it what you will. Everyone I know loves it, but that's the crowd I hang out with *(laughing continues)*. . . . But, you know, something that people miss—that in the heat of that moment, if you will, is easy to miss—is that the manifesto also begins immediately with the figure of the immune system.

DH: Yes, it does.

CW: And it begins with the question of race.

DH: And conquest. It's absolutely about inheriting the histories of indigeneity and race.


## BIOPOLITICAL WORLDINGS

CW: That's right. And so to put it this way, at this moment in the history of the kind of work that we do, to talk about race and to talk about immunity is to automatically be in a biopolitical discourse. And when you remember that the fundamental logic of the immunitary mechanism in biopolitics is essentially the logic of the *phar-*

*makon,* what are we back to? We're actually back to irony. We're back to a retooling and exfoliation of what was going on with the term *irony* in the "Cyborg Manifesto."

DH: And the whole question of emergent natures/cultures in the "Companion Species Manifesto" is about the dilemma of inheritance, of what have we inherited, in our flesh.

CW: Right.

DH: And, you understand, "Ms. Cayenne Pepper continues to colonize all my cells—a sure case of what the biologist Lynn Margulis calls symbiogenesis."

CW: Right.

DH: So if the "Cyborg Manifesto" is looking at the couplings of cybernetic systems and organisms, the "Companion Species Manifesto" is saying, "Wait a minute, the entity that we are is the outcome of a symbiogenetic doing." We are sympoietic systems; we become-with, relentlessly. There is no becoming, there is only becoming-with.

CW: Right. And it's also a logic of both/and, as irony was in the "Cyborg Manifesto." The way I think of it is that if you take the opening that irony made possible in the "Cyborg Manifesto" and then you make it fleshy, you make it evolutionary . . .

DH: Yes. And note, Cayenne and I, in the most literal sense of the term, before you get to the second paragraph of this little quote, look how we are linked in becoming-with each other: it's clear that one of us has a microchip injected under her neck skin for identification, one has a photo ID California driver's license. We are subject to state regulatory identification apparatuses and biopolitical identification apparatuses and surveillance; the microchip injected under the neck skin is a direct thread to the "Cyborg Manifesto." I remember deliberately writing it that way. So we're working from the beginning within the biological/biopolitical discourse of canid/hominid, pet/professor, bitch/woman, animal/human, athlete/handler: the questions of these multiple configurations of who and what we are in a Foucauldian sense of discourses, discourse production. The material semiotic ferocity of that.

cw: That reminded me of a great piece by the bioartist Eduardo Kac. He injects an animal identification microchip under his skin, but as the piece gradually takes different shapes, what you find out is that he has Jewish relatives who were killed in the camps.

DH: Whoops.

cw: Exactly. And so something that started out as one type of piece . . .

DH: . . . becomes quite another . . .

CW: . . . type of piece—talking about driver's licenses, forms of identification and surveillance, and that sort of thing.

DH: Right, and this does, too. It looks like it's just sort of a light
joke. But pretty quickly, because she's a U.S. herding dog, since
her ancestors are the dogs who worked to develop the agrobusi-
ness ranching practices of the U.S. West after the Gold Rush,
she and I are children of conquest. From the beginning.

CW: Right. Unambiguously so.

DH: From the beginning. And the question of whiteness is right
there from the get-go, the question of living on both Native and
Californian land, palimpsestic layerings of sovereignties, are
there from the get-go. So that deep kiss is quite literally a deep
kiss.

CW: Oh yes.


SIDEWINDING SYMPOIESIS: MAKING KIN


DH: And it's deliberately nongenital—not only did we actually
do the kind of kissing so described, and we didn't do anything
else *(laughing)*—but in addition, it is deliberately—now I am
speaking as a biologist, one of the voices of that book—this kiss
was deliberately about lateral transfer. It was a commentary
about the tree-based lineages that overemphasize the notions

of DNA, linear transmission of DNA as "the book of life," and from the get-go that little soft-porn statement is saying, "Nonsense." You know, the first name mentioned is Lynn Margulis's.

cw: One of the fascinating things about the book is to step back and remember how it begins and ends. Everybody remembers the kiss, but the book ends with the hilarious scene between Cayenne and Willem: 35-lb. Cayenne the Aussie and 110-lb. Willem the Great Pyr having this utterly nonreproductive, dysfunctional, and funny sexual intimacy. So one of the strong and easy-to-be-missed—and yet very biopolitical—aspects of the book is that the book begins and ends with nonreproductive sex.

DH: And on purpose! And with Norman O. Brown's *Love's Body*, and with Charis Thompson's term *ontological choreography*, in the historical conjuncture where who "we" are, whoever this "we" is, in this thick now, ontological choreography is both what makes us who/what we are and also what we must engage. We must engage—must dance—ontological choreography if we are to live and die well with each other in the troubles. For many reasons, some of them in the "Companion Species Manifesto," my slogan these days is "Make Kin Not Babies!"

cw: One of the things that that final scene—

DH: I'm glad you find it funny. I still find it hilarious!

*Companions in Conversation*

cw: Oh yes, I think it's incredibly funny, and it reminded me of—and every time I've used this example with students, it always takes them a second to get to the existential biological math—maybe my favorite line by Vicky Hearne, from an essay (I think) called "A Walk with Washo: How Far Can We Go." Vicky says the surest sign of how intelligent Washo was occurred when she saw her in a tree masturbating to a copy of *Playgirl* magazine. *(Both start laughing.)* You can sort of see the wheels turning with the students when you tell them this *(more laughter)*. But it's what would become the great "scandal," which now is not news but was then in biology.

DH: It's still news, alas . . .

cw: . . . of nonreproductive sex, and how that becomes a gateway to these other, much more complex phenomenological lifeworlds that do and don't overlap between different species.

DH: People had gotten used to, way too easily, concepts like *aggression* and *competition* being used with other critters, as if they were technical terms, just as if they weren't extraordinary anthropomorphisms, but would react very badly if questions of desire or labor or friendship were raised. That's a passage about desire, and it's not about us.

cw: No. And this is where I think *play* actually circles back to the opening of the book and is so important—not just in your work but as a general topic for further investigation—because a major problem in biopolitical thought has been that it essentially becomes what a lot of people have called a discourse of "thanatopolitics"; it becomes a discourse of the body as "bare life" being exposed directly to violence.

DH: Forced life, forced death . . .

## AFFIRMATIVE BIOPOLITICS
## AND FINITUDE

CW: Right, and so what a lot of people have tried to think about is what an affirmative biopolitics would look like, to use Esposito's term. But the problem with a lot of those efforts has been to typically fall back into a kind of uncritical affirmation of "Life," capital *L*. It's as if you took Foucault's famous statement "Resistance is on the side of life," and you put it on an LSD tab and handed it to everybody *(both laughing)* . . .

DH: . . . and simultaneously dressed as the Borg queen and wore a placard saying, "Resistance is futile" *(still laughing)* . . .

CW: Right, and so (in Esposito's case), working in an Italian, Catholic context as a philosopher of "Life," you see what I mean. One of the important ramifications of the "Companion Species Manifesto" is very much its contribution to the question "What would an affirmative biopolitics look like that was not simply an uncritical affirmation of 'Life,' capital *L*?" as this kind of flat ontology of positively valued concatenations of *élan vital*. And so the importance, not just of play, as the book opens, but also of joy.

DH: And also of questions of shared authority, the training chapters in there, the questions of nonmimetically experienced suffering and achievement. The "Companion Species Manifesto" doesn't deal with questions of dying nearly as much as the stuff I'm writing now does, partly because of the inescapability of needing to think better about extinctions.

CW: Right . . .

DH: Extinctions and exterminations and genocides. The "Companion Species Manifesto" is not fundamentally a work that deeply inhabits those biopolitical matters. But an affirmative biopolitics cannot be a pro–Life politics.

CW: No, no, not in—

DH: In the United States context, folks react immediately when you say, "I am *not* a pro–Life thinker." The resonance of the abortion struggles is of course immediate and on purpose. But I think an affirmative biopolitics is about finitude, and about living and dying better, living and dying well, and nurturing and killing best we can, in a kind of openness to relentless failing. I am a resolute, non–pro-Life feminist. And affirmative biopolitics is probably a pretty good phrase for that, but it won't work

*Companions in Conversation*

as a slogan! There was a pretty good slogan for the "Cyborg Manifesto" that Elizabeth Bird came up with—"Cyborgs for Earthly Survival." But I don't have a comparable slogan for what I agree is the affirmative biopolitics of the "Companion Species Manifesto" that insists it's about being mortal and finite together in our absolutely nonmimetic difference. It is about significant otherness. Maybe it can be the slogan from a sticker the ecosexual artist Beth Stephens and her spouse, Annie Sprinkle, gave me: "Composting is so hot!"

## FORCED LIFE, DOUBLE DEATH, HOLOCAUST

cw: I think this is a place where there is a crossing, or at least a friction, a rubbing up against, between ecological thinking (you mentioned extinction earlier), what you might call an ecopolitics, and biopolitics. Because one of the interesting problems that you've talked about, and Derrida has talked about, and a lot of people have talked about, is parsing the differences between these terms *killing*, *death*, and *letting die*.

DH: Right.

cw: And the extent to which these are and are not the same kind of violence.

DH: And then the terrible violence of making live. Eric Stanley, who did a dissertation in History of Consciousness, was particularly ferocious. He's a very strong pro–animal person, a subtle and wonderful thinker, who made me think much more about the violence of making live when the possibility of living well is actively blocked. The vast machines of forced life for purposes of extracting value, for purposes of slaughter. The multiple forced-life machines are perhaps the greatest source of violence on our planet, if one's talking about the other critters. And the forced life of the supermax prisons, too. The multiple machines of forced life that Eric wrote about—finding necropolitics insufficient for thinking about the problems of biopolitics.

CW: Right. I wonder if in that light—and this is something I wanted to ask you anyway—if you would change a little bit the view that you expressed in the "Companion Species Manifesto," about this *figure* (if that's the word we want here, and who knows if that's the word we want?) that a lot of people have used, not least of all Derrida, of a "holocaust," actually, of nonhuman life that some people have associated with the sixth great extinction event of the planet, and others, like Derrida, have used to talk about the killing of nine billion animals per year in North America for food.

DH: Probably more than that. If you consider invertebrates and fish, it's way more than that.

*Companions in Conversation*

cw: A whole lot of animals are not included in that number. Of course, Derrida is quite aware of the complexities he's walking into here, being an Algerian Jew, which is interesting in itself. The whole status of Jewishness and of "writing as a Jew" is very interesting and complex . . .

DH: Right, and his never having been properly "human" within the Western philosophical canon.

cw: That's right.

DH: Jews have never been "properly" human.

cw: That's right, and so I wonder if that analogy still works—because you pretty stridently come out against it, for reasons that I understand, in the essay . . .

DH: I would do a lot of that differently if I were writing it now. For one thing, Marco and I would not be heading out to Burger King before our training sessions at the animal shelter! *(Laughter.)* I've changed in my politics in relationship to the questions of industrial animal agriculture. And I am still not a pro–Life activist. I think that's an exterminationist position. I think the question of working lives, including killing for food and killing for market, remains potent and necessary. Besides that, working animals matter; their actual work deserves respect. I think I am engaged in an affirmative biopolitics. Well, there's a long conversation to be had there, but I have definitely, without question, moved from 2003 on these matters. And the question of the animal holocaust and the questions of animal geno- cide . . . first of all . . .

cw: . . . both through killing and through letting die . . .

DH: . . . and forcing to live . . .

cw: . . . yes, making live . . .

DH: . . . making live in vast numbers in order to kill. Making live in appalling conditions in order to kill in appalling conditions — for profit. The question of capitalism cannot be left out of this.

cw: No.

DH: *And* the question of the still-ongoing vast expansion of the human population, and what counts as wealth in this human population. So I have changed since I wrote that stuff. I've al- ways—then and now, but even more now—felt that we need more than one word at a time, and we need to be careful how we situate these words in relation to each other. When I talk and write about these things, I follow Deborah Bird Rose and Thom van Dooren and the other Australian Extinction Studies people who ask what it means to live in a time of exterminations and extinctions. And I've added to that multispecies, human and nonhuman, genocides. What is it to live—in extended time, you can date these things variously, but it's not an arbitrary matter — in an extended time of extraordinary surplus killing and surplus dying? In what Debbie Bird Rose calls times of "double death"? Death is not the problem, but cutting of the tissue of ongoingness is the problem. What is it truly to live responsibly in times of exterminations, extinctions, and genocides? Add holocausts to that list. And add extraordinary increasing human burdens and numbers. And I would add that the human–animal divide does not sort into those words; you can't sort humans and animals into piles with those different words; this is a multispecies affair.

cw: No, and the biopolitical point is that those kinds of species distinctions are not constitutive of the problematic.

DH: They're not constitutive of the problematic, which does not mean that the specificities of different critters don't matter in making judgments.

cw: Right. They matter all the time.

DH: That's exactly what matters—the concretenesses are exactly what matter. So yes, I would write differently about "holocaust" and all its kin. I engage a little bit in the "Companion Species Manifesto" in the problems of factory ranching, the Animal Industrial Complex, but I would be stronger in that part of the "Companion Species Manifesto" if I were writing it now. And remember how strong Barbara Noske and Carol Adams and others were twenty years before I wrote about any of this.

cw: We're back at this point to what I think is a very well-articulated commitment of yours, not just here but in your other work of this period, that looks forward to the biopolitical context, and the designation by race or species as making killable but not murderable. And here, I think you and Derrida, perversely enough, are the perfect couple, because the point for both of you is that the ultimate fantasy is to think that you can step outside this violence that you're implicated in . . .

DH: . . . and neither of us thinks that you can . . .

cw: And so the question does not become just "killable but not murderable," and it does not become just "Thou shalt not kill," but it becomes, as you put it, "Thou shalt not make killable." It's on that specific terrain that I think there is an opening that has yet to be fully worked through, a crossing between biopolitical thought and ecological thought, because part of what animates your work in light of that commitment is to say, "Look, if the issue is 'Thou shalt not make killable,' then it's not about escaping killing or escaping death. It's about what posture or what stance does one take toward life."

DH: Toward *this* killing . . .

cw: Toward *this* killing or this life in its specificity . . .

DH: Toward *this* living and dying, *this* nurturing and killing . . .

cw: . . . in its specificity . . .

DH: . . . in its more-than-doubleness . . . .

*Companions in Conversation*

cw: That's right. And so it seems to me that one of the ways this question expands in many different directions, in terms of the crossings of biopolitics and ecological thinking, is also by bringing into the conversation the question of "letting die," because a point that you've made, and that Derrida makes very strongly, is that "our" entire way of "life" is predicated on the violence of a *massive* "letting die"—not a direct killing, not an execution, but a truly massive letting die alongside of practices that are quite clearly making killable but not murderable, like factory farming.

DH: And alongside an apparatus to "make immortal" a small fraction of the human population, if possible . . .

cw: . . . right, ever smaller and ever more "immortal" . . .

cw: . . . through whatever fabulous, fantastic techno-fix—that actually has shaped too much of our medical system. But let me say something about this another way. This is a very complex nexus of questions . . . .

DH: You're not just kidding. You pick up one thread and you're aware of six you just dropped. But this crossing of the biopolitical and the ecological—another place, it seems to me, that this is in our conversation right now around making killable has to do with questions of species recovery plans and habitat regeneration plans and various ways of engaging in ecological res-toration, ecological recovery, species recovery, and so on—these very complicated and never innocent, very important but also very fraught engagements. There was a talk at the confer-ence this past weekend around a dilemma involving some of the islands off the coast of California, where removal of so-called *invasive species*—itself a term that "makes killable," the very use of the term *invasive species* makes killable, whether you're talking about immigrants from Central America or rats and cats on an island. "Invasive species" is, literally, a powerful way to make killable. So consider an island world, where ground-nest-ing birds and many other critters cannot continue the tissues of their ongoingness and are undone by these rats and cats. Some people just want to call them "species out of place." Well, that's true, there's a truth in that. And there are a lot of things you can call them, there are lots of euphemisms: "species removal" . . .

cw: . . . "species relocation" . . .

DH: . . . and I said to this really sensitive biologist/ecologist who gave this talk, "Well, look, this seems to me very similar to the question of a woman who knows that she is pregnant and can-not carry the child to term, where she knows she is killing. Why do we pretend to ourselves that this is not an extirpation, a killing? What sort of innocence is this? What does it enable not to know/admit that we are killing? The being, human or not, should not be made killable, *and* killing is sometimes the most responsible to do, is a good thing to do, even—but never an innocent thing to do." How can we really live in noninnocence, because I really think we must?

I don't think we have a chance to live responsibly insofar as we are pro–Life. The search for innocence is exterminationist. I think we need pro–ongoingness in our mortality, not pro–Life. And judgments are made about that island ecosystem—judgments that are flawed and historically specific, and for some critters and not others, and for some people and not others. And killing ensues. Why not admit, "I am in fact going to engage in deliberate killing. . . . Every rat and every cat on this island I will kill to the point of local extermination, and I will not name it with euphemisms or dress it up with an excuse, and I will still say this is what I *should* be doing, and simultaneously, I am *not* innocent. These killings, these deaths, these particular critters, matter." That's a little bit of what I mean by "not making killable"—that in order to be *for* some ways of living and dying and not others, in order to be *for* the ground-nesting birds (in this example), in order to be *for* the partial recovery of this island's plant, animal, and microbial ecosystems, I/we must kill. But I'm not going to hide behind terms like *invasive species*. I am not going to make killable. I am going to argue *for* this worlding in an interrogative way that asks, Is this "us," is this a "we" that we will cast our lot with? Or not?

CW: And that brings us back in a really unsettling way to the questions we were talking about earlier around *human* population. Because a pointed way—and you know that this goes back to debates that have been going on since the '60s and '70s—to put the question would be simply to ask, "Would you say exactly the same thing about members of the species *Homo sapiens*?"

DH: Why would we want to? The devil in the situated details.

CW: Precisely so, because if indeed within the biopolitical problematic—or the ecological problematic—species isn't constitutive of the problematic, then the first hand that has to go up is, "Uh, well, if we want to talk about—let's just call them 'destructive species'—then we need to start with . . ."

DH: "Let's go to the top of the current list," which is exactly what made it "the Anthropocene." The Anthropocene gets its name from making that the head of the list. . . . The Anthropos is the destructive "species"—Man, once again, the "species."

CW: Right.

DH: And that's also what's wrong with the figure of the Anthropos. It's not a "species act"; we're not doing this as a "species." What is happening that gets called the *Anthropocene* is a situated complex historical web of actions—and it could be, could

*Companions in Conversation*

have been, otherwise. But people forget that, partly because of the power of the word. People really believe that the human species is doing this thing, as an act of human nature. And it's simply empirically not true.

cw: Well, that's the funny thing about the term. I participated in this huge event in Berlin on the Anthropocene last year—this huge, hypercurated, European thing. It was great because a lot of people who were there were great, but one of the things that came into focus was precisely this problem about the term that we talked about this morning: that for half the people it's the ultimate posthumanist term—in the sense of utterly decentering.

DH: In some ways it is.

cw: And for the other half of the people it cuts exactly the way you're describing.

DH: And then for the *other* half of the people, since we have many halves . . . As I said this morning, I don't have to choose just one term. If I did, it would be *Capitalocene,* and that figures the subject differently. And it's at least as interesting for those for whom it might be the ultimate posthumanist term—*Capitalocene,* I mean. It could satisfy some of the same needs, but it will cause different troubles. Response to the Capitalocene demands systemic change located in flesh-and-blood, situated, complex histories.

*Companions in Conversation*

cw: Well, it certainly doesn't take a single species and bring it front and center, knowing Marx's analysis.

dh: Capitalism isn't just a species act. And Capitalocene asks something else. Capitalism obviously isn't just one thing. It's obviously a very complicated *historical system* phenomenon, among other things; it has many histories and unevennesses in time and space. And you can't date it from the middle of the eighteenth century with the steam engine. The plantation system is surely more fundamental; it is ongoing, too (think of current oil palm plantations and associated destruction of mixed forests and their human and nonhuman lifeways). You cannot run the debate about what the Anthropocene means between the deep ecologists, on the one hand, who put it with the invention of agriculture, or even with Pleistocene human hunters, or the appearance of *Homo sapiens sapiens* on the planet, or something, and, on the other hand, fossil fuel–using humanity with the internal combustion engine, and following.

The Marxist political ecologist Jason Moore does a nice job of getting us started on this. You cannot even begin to think the complexity of capitalism as this earth-making thing without going to the trade zones in the Indian Ocean in the fifteenth century, the many world-making trading zones and wealth accumulation zones and inventions of plantation agriculture, and the moving of plants and animals and microbes and people around, and the deforesting of the river basins in the sixteenth century. You cannot even begin to touch this question with the binary time problem that emerges almost inevitably when people talk about the Anthropocene. Either you're talking about the past two hundred years, or you are talking about, you know, the dawn of the species. And then you get this fight between the deep ecology–oriented people and the folks who are worried only about the fossil fuel economy. This will not do. The complexities of time and space are ill done that way. And Capitalocene does a better job on that point. *And* it asks which populations of animals, plants, and peoples—*and* microbes (since, let's face it, the questions of fermentation and disease are fundamental in the history of capital, big time—tell me about WWII without fermentation!)—anyway, the players in Capitalocene are, at a *minimum,* situated plants, animals, humans, microbes, the multiple layerings of technologies in and among all this. If you think the Capitalocene, even in a remotely smart way, you're in a whole different cast of characters compared to the Anthropocene.

cw: One thing that we talked about earlier, that I do think is of interest in the term *Anthropocene*—and this actually brings us back to the "Companion Species Manifesto"—is that everything we've just said invokes not just, on some level, the radically ahuman and unthinkable time scales of geological time that are invoked by the Anthropocene, but also the temporal asynchroncities that obtain (and you can think of those generationally or however you want) be-

*Companions in Conversation*

tween these different kinds of biological and technical forces and their developments, that are stitched together to create what you are calling Capitalocene.

DH: You can speak those things in the Anthropocene, but you *must* speak them in the Capitalocene.

CW: Right.

DH: Those asynchronous and distributed over time and space forces and complex system-property ways . . .

CW: That's one reason we're not back to . . . the Sublime!

DH: We sure aren't! *(Both laughing.)* Partly, I think, the *Anthropocene*—for various reasons, good and bad—the term got popular, and it got popular with the scientists, too, and it got popular with the geologists. And, mind you, it's worth remembering that the people who proposed this term, which is only around the year 2000, the first person to propose the term is a biologist in the Great Lakes who studies freshwater diatoms, right? And he's looking at biogeochemical processes. The whole term *ecosystem*, the very word *ecosystem* also comes out of freshwater lake ecology, and grasslands ecology, too, which I think is interesting. And it comes out of biogeochemical processes, in part through the linkage with the Russian biologists who *invent* the term *symbiogenesis* in 1910. The Russian biogeochemists in my life come to me through my thesis adviser G. Evelyn Hutchinson.

*Companions in Conversation*

Anyway, what I'm signaling here is that the biologist who invented the term *Anthropocene* and then joins (no surprise here) with an atmospheric chemist—they are worried about the bleaching of the coral reefs from heat and acidification. Their focus is very earthy, very fleshy, very much about biopolitics, ecopolitics, extinction, all that. And others also worry about the bleaching being possibly partly due to a bacterial *vibrio* infection, an infectious event with the same group of critters (bacteria) that cause cholera, on the one hand, and that are involved in developmental signaling in some ecological evolutionary developmental biology symbioses. Eugene Stoermer and Paul Crutzen focus on the bleached coral reefs, and they're consumed by the anthropogenic, human-caused processes that are written in the rocks, the waters, and the atmospheres. And the geophysical unions form working groups to figure out if the stratigraphic evidence is sufficient to rename the epoch by straight-up standards of their profession. Is the Anthropocene a boundary event, like the K-Pg boundary separating the Cretaceous from the Paleogene Period (Scott Gilbert's idea), or is it an epoch or an even bigger geohistorical category?

This is a conversation we need to be having. So I don't want to toss out the baby with the bathwater, you know; I wish that because the term *Anthropocene* carries so much else besides what they intended, I wish that it hadn't been their term. But it is their term, and I hope that when the working group makes its report in 2016 the geologists adopt *Anthropocene* as an official term. We will need to continue to operate within this discursive materiality as well as others that name our urgencies better in key respects. I think Capitalocene should have strong discursive materiality among us, but there are no institutions to do that. You can't even talk about capitalism in the United States. You can't say the word without being read the riot act—I mean truly, it's an unspeakable word most places!

CW: This goes back . . .

DH: Even if you're a capitalist, you can't say the word *capitalist*. *(Both laughing.)*

CW: Right, right, you're just talking about "economics," you know, as if that's taken for granted.

DH: My friend Chris Connery points out to me that the Chinese talk about capitalism all the time (and the ongoing Cold War). And if ever anybody was in the middle of a capitalist revolution, it's China. In a lot of ways, even with less formal freedom of expression, everyday political talk among the Chinese is *way* richer than among us (even though in other ways, it's not).

CW: Which wouldn't be hard! *(Laughs.)*

DH: Well, it wouldn't be hard. But on the other hand, the killing of the Left in China—Left discourse in China—is tragic. But then there's little enough Left discourse here either. Anyway, though, enough about capitalism for now.

## *Companions in Conversation*

cw: I wanted to come back to something, maybe slightly less depressing, but also on a smaller scale, that has to do with talking about the biopolitical dimensions of the "Companion Species Manifesto." It's about joy and play and about quality of attention, and the kinds of responsibilities that that involves with the creatures in question. What I'm thinking of—you mentioned this to me several years ago, and it comes up in the "Companion Species Manifesto"— is the passage about the "metaretrievers." (DH *laughs.*) You write Vicki Hearne this letter, and you're talking about metaretrievers and taking Cayenne down to the beach.

DH: Roland and Cayenne both.

cw: This activity that you recount would now be, as I understand it, illegal.

DH: But practiced.

cw: Illegal but practiced. And so here is the question I want to ask. You talk about how the term *companion animal* emerges from a very historically identifiable complex of medicalization and academic institutional life. And there are other cognate terms that we could think of, that we probably don't like a whole lot, that come out of this same biomedical, biopolitical context (as Foucault would call it). What I want to think with you about—and your work gives us a vocabulary to do this—is how this kind of increasing regulation and medicalization of how human and nonhuman creatures interact (all in the name of enhancement and security, of course) is in fact part of a much larger biopolitical fabric. And to make sense of this, we have to change what we think "politics" is. To come back to the beginning of my question, one thing you do in the manifesto—both in how you start it and how you end it, but in other ways, too—is to make it clear that these aren't just theoretical questions about what biopolitics is: these are part of the same fabric of life . . .

DH: . . . these are very ordinary . . .

CW: . . . and mundane, but also very death-by-a-thousand-cuts . . .

DH: . . . but also joy-by-a-thousand-cuts . . .

CW: . . . at the same time, because the flip side is also true: your dogs have literally historically unprecedented access.

DH: And unprecedented wealth.

CW: They have access to forms of veterinary care and quality of food and all sorts of things.

DH: Our chicken coop is bigger than the shanties that many people are living in, in the megacities of the world. Or tents for refugees in war zones. I know this . . .

CW: And so one thing I'm interested in on an even deeper level with these developments—and this does take us back to our discussion at the beginning of the manifesto and its beginning with the immunitary paradigm . . .

DH: . . . both inheritance and the immunity system . . .

CW: . . . is that all of this seems to be part of the fabric in which, at least at Rice—I don't know how it is at Santa Cruz—outside of every elevator on campus there are Purell dispensers mounted on the wall . . .

DH: . . . and now they say they are "antibiotic free" and "alcohol only." There's a whole discourse evolved in those little pumps.

CW: Right. So we have that, and we have a crazy number of signs that I see that say, "Please pick up after your dog"—not because nobody likes to step in dog shit but because "dog waste transmits disease." And at the same time we have this explosion of things like food allergies and various kinds of autoimmune disorders . . .

DH: . . . and a completely epidemic-friendly global industrial food system.

CW: And so what the "Companion Species Manifesto" does—and in ways that I do think inherit some of the work of the "Cyborg Manifesto"—is to put together a vocabulary for helping us to understand that, look, these aren't just little embarrassing or annoying "ethical," "lifestyle" issues; these are actually part of a larger political seismic shift—in the name of "making live," in the name of "enhancement," in the name of "security"—

DH: This is part of biopolitics . . .

cw: ... that is actually enfeebling ...

DH: ... and are provocations to thinking.

cw: Yes. I wonder sometimes. I think back on kids of our generation, and I think, well, maybe we were better off without this enhancement. I tell my students, "You need to eat more dirt!"

DH: Anna Tsing, whom you know is a very close friend and who-I-want-to-be-when-I-grow-up kind of colleague ...

cw: Yes, she must be incredible to talk with about this stuff.

DH: ... she told a little story the other night when we were getting ready to have dinner and watch bad TV—we were talking about dogs, and we were talking about Cayenne's habits. Since we're in a drought, we're not flushing the toilet very often. Her habits of drinking out of the toilet bowl have become a bit of a household problem.

cw: *(Both laughing.)* "Come give Mommy a kiss!"

DH: We were laughing about the sheer materiality of living together, just the sheer thisness of it, and the absurdity of it and the forgiving each other for stuff. And Anna says, "Well, I did my fieldwork originally on the island of Borneo among a group of people who were agriculturalists during that period—deforestation has since undermined their ways of making a living significantly—they traded a lot with the local Muslim populations.

*Companions in Conversation*

247

They were a very complex part of a very Indonesian fabric." And she says, "But, you know, in fact there were a lot dogs around, and the dogs were in and out of the houses, and they're pretty much living independent dog lives, but fairly closely associated with the people, too. The dogs hung out with kids a lot. You know," Anna said, "the dogs were the diapers. The dogs ate the baby shit. And this was absolutely expected of them. It helped keep the houses clean. Nobody had access to cotton diapers, much less synthetic-fiber diapers. The dogs obviously totally enjoyed it, the babies clearly loved it, and none of the adults seemed to think there was the slightest thing wrong with it—quite the opposite. The dogs ate baby shit." Now I guarantee that this is not going to be a popular way to deal with the diaper waste in the landfills in the United States *(laughing)*. But it does kind of make one recontextualize these questions of sanitation, security, waste, and biopolitics and biotechnologies.

cw: The overhygienization—and in biopolitical thought, this takes us all the way back to Foucault: the overhygienization and . . .

DH: . . . and the misunderstanding of historical multispecies life . . .

cw: . . . within an immunitary and autoiummunitary context, all in the name of a form of well-being modeled on class- and race-based notions.

DH: All in the name of well-being. The apparatus of biomedicine, and the apparatus of immunology in microbiology, grew up in a framework, in a colonial institutional framework, of getting rid of the enemy and managing the subordinate. Sterilization, exclusion, extermination, transportation, so on and so forth. Biomedicine did not grow up in "How do you cultivate assemblages that maintain multispecies, culturally diverse, ecosystem health?" There is a truly tectonic shift going on these days in biology and medicine and microbiology across the world —unevenly, way too slowly—but no critter on this planet is left out of this—in some really deep ways that are affecting experimental practice, clinical practice, and so on—critters are understood to *be* ecosystems. If you're serious about enhancing the health of some ecosystems rather than others, you've got to think in an ecosystemic way. Which associates/companions should be here, and which should not? Which critters are always disease causing in an extremely serious way, where we really need to find ways of excluding those guys, and which other guys are actually really good at excluding the ones you don't want? And so forth. Because you literally can't sterilize; the hand-sanitizer thing is a bad joke. The main point is that insofar as biopolitics is concerned, this question of ecosystem assemblages is the name of the game of life on Earth. Period. There is no other game. There are no individuals plus environments.

*Companions in Conversation*

There are only webbed ecosystems made of variously configured, historically dynamic contact zones. With the help of my colleague-friends Karen Barad and Scott Gilbert, sometimes I name this intra-active and diffractive complexity GeoEcoEvo-DevoHistoTechnoPsycho sympoiesis! The series expands and folds back on itself . . .

CW: . . . as the former Santa Cruz professor Gregory Bateson reminds us. I do think it's in this context that one can excavate a deep line of connection between the two manifestos, terms like *companion species* (and some of the other terms we've been talking about), and essays like "Sex, Mind, and Profit," "Situated Knowledges," your essay on the immune system, "Promises of Monsters," "Virtual Speculum," and so on. In a way, those essays plus the two manifestos become a kind of virtual book in their own right. Their lines of connection to the "Companion Species Manifesto" are a little harder to recognize just because the surface of the text is so different, but in a fundamental sense, the underlying theoretical dynamics that connect them . . .

DH: . . . they're deeply braided . . .

CW: . . . haven't fundamentally changed. It's how they ramify these issues in different ways.

DH: They both tell technology stories, evolution stories. They both tell stories of intimacy and pleasure. Both manifestos are engaged in all these forms of storytelling, but the balance is different, the foreground/background is different, the genre is different.

cw: And I think the stakes are different in the sense that the sites on which those same dynamics play out are different. That's why I loved the way that the "Companion Species Manifesto" ends with the fascinating discussion of the Sato street dogs in Puerto Rico.

DH: Yes.

cw: That's a great example of how we may be talking about "the same" theoretical dynamics, but how they ramify geopolitically and culturally is very different.

DH: This was an attempt to emphasize questions of race and national power in dogland. And many other things. And also make plain the diverse biopolitics of humans and dogs. I was fascinated in the *Sato* dog story in Puerto Rico by accounts of street people taking care of street dogs. Also care by people who weren't street people, but still living hand to mouth. Their practices of relationship with street dogs deserve to be foregrounded, storied, protected, and respected. The image of the "forever home" in Massachusetts, and always being "rescued"—really "rescue" discourse took over "adoption" discourse in dog life in the United States, too—is very problematic. That every good dog is a rescued dog is, among things, a colonialist discourse. I wanted the *Sato* piece to complexify this particular international adoption story, without making international adoption, of dogs or people, the enemy.

cw: And that happens, those kinds of pressure points emerge constantly, even closer to home. We were talking earlier about Katrina and Houston and New Orleans; those issues came up around the discursive status of the "refugees" from New Orleans coming to Houston.... Well, hang on a minute, maybe we need to think about a more responsible way to talk about what's going on here and the position these people are in and what they've been through.

DH: Among other things, being called "refugees" makes them not-citizens.

cw: Exactly.

DH: And of course, in very significant ways, the way the rebuilding of New Orleans has proceeded has continued to exclude them as much as possible.


THE PRACTICE OF JOY: MAKING KIN

DH: I want to go back to two things that we touched on before we go any further. One of them has to do with the question of joy, and the importance of the practice of joy in living our mortality with each other. If we are to develop political vision, if we are to develop some sense of living and dying with each other responsibly, including responsibly to "the troubles," I think the practice of joy is critical. And play is part of it. I think that engaging and living with each other in these attentive ways that elaborate capacities in each other produces joy. In the conference on Arts of Living on a Damaged Planet, Deborah Bird Rose called this thing "the bling of life," and then she called it "shimmer." She was talking about the way some of her teachers, in particular her Aboriginal teachers in Australia, called it "shimmer" (that was the best way she could translate it). And she was thinking about this in relation to the question of the flying foxes in Australia and their fruit trees, and the obvious sensual pleasures of the flowers and the bats and how they move toward each other, in what Natasha Myers and Carla Hustak call this sensual loquacious involutionary momentum of life. This is a biological discourse, among other things. And I think it's really important to participate in this bling of life—to be able to be attentive to and be able to describe the shimmer.

cw: As we were discussing earlier, I think this is a crucial resource for the specific juncture that biopolitical thought is at right now, which is trying to say what an affirmative biopolitics would look like. And I think part of the problem, for work in "theory," is that when you start talking about things like joy and play, you're marked as if you're automatically not talking about politics in some serious way.

*Companions in Conversation*

DH: And I think quite the opposite. I don't think we can even begin to understand what it takes to be political in these times without this.

CW: No—and I think this is also a deep point of connection between your earlier work and the "Companion Species Manifesto." To me this is a huge inheritance and resource and legacy from feminism, and aspects of the women's movement and feminism that *were* about joy and *were* about affirmation . . .

DH: . . . and queer politics. . . . And never only domination.

CW: Yes, and I want to talk about queerness in a moment.

DH: I think that the politics of pleasure were thought, developed, practiced, proposed as *public* practice most vividly in queer movements.

CW: I don't know what your feeling is about this, but I felt that when you wrote the "Companion Species Manifesto," one of the reasons you moved away from the figure of the cyborg . . .

DH: . . . they're in the same litter . . .

CW: . . . yes, they're in the same litter, but as you suggest in "The Companion Species Manifesto," you opted for this other figure of the companion species because the figure of the cyborg was not queer enough for the work you wanted to do then.

DH: Yes, I think that's true.

*Companions in Conversation*

CW: It may have been "theoretically" queer . . .

DH: And also not intimate enough, though plenty intimate . . .

CW: But you put your finger on it a second ago when you used the word *pleasure*.

DH: Yes.

CW: It's the connection between queerness and pleasure—which moves us into an affective register that draws on the nonreproductive sex and intimacy and joy with which the book begins and ends.

DH: The nonheteronormative, as it got labeled in really an unfortunate combination of syllables *(laughs)*. . . . Now, repeating myself, I just say, Make Kin Not Babies!

CW: Those are resources for doing two things: not just thinking affirmative biopolitics but—to go back earlier in our conversation—thinking about kinds of ethics and politics that have typically been taken as not having a lot to do with each other, namely, biopolitics and ecological thinking. Part of it is that ecological thinking, up until pretty recently (thanks to people like you, but there are other people doing this work) has often smuggled back in—even in the name of biodiversity—a lot of the reproductive discourses that are not queer enough.

DH: That's all certainly true, although I also need to say, "Yes, but." I think of ecological politics, from as early as you want to get—with Val Plumwood, for example. Ecological politics for me doesn't do that, through and including Thom van Dooren today.

CW: Right, but the lineage you're invoking is precisely the one that's needed.

DH: Of course, it's not an adequately well-known lineage, but it deserves to be. If a mainstream journalist is going to write his or her story and they want a little backstory, they're not going to know *that*, but they need to.

## TELLING STORIES, CULTIVATING RESPONSE-ABILITY

DH: Here's another little story I wanted to tell while we're still on biopolitics—and then I want to go to religion, actually, or wherever we go next, but we need to get to religion. But we were talking about the huge change in the way that we live with our cats and dogs and parrots and chickens and whatever (and understanding that having a parrot is already a very iffy thing to be doing ecologically and in other ways). Anyhow, the point is that we live with our critters differently. They are family members, kin, in a way that was not imaginable even in the 1950s, with Lassie et al. And of course, for better and worse, they acquire not just the right to health but the obligation to health, a very dubious acquisition! They are within biopolitics, like it or not.

cw: Compulsory making live.

DH: Compulsory making live. The apparatus of biomedicine, pet medicine—they are in this apparatus big time. And it's pretty expensive. But I often think in terms of little stories or tiny details or tripping over something that opens up into huge worlds, where thread by thread by thread, as you spin from some tiny thing, you are relooping together the worlds that are required for living and dying *here,* with these details. Okay, so I'm living with an elder dog now, Cayenne, and one of the things that happens to older spayed female dogs (and postmenopausal women) is that their urinary sphincters become lax, and they begin leaking. And there's a drug, phenylpropanolamine, that is routinely prescribed that works pretty well to tighten up their sphincters so that they can continue to live indoors and, among other things, sleep on the bed, because nobody wants to sleep in the wet spot.

But phenylpropanolamine (PPA), an angiotensin, tends to increase blood pressure. Cayenne developed a heart-valve problem that made increasing peripheral blood pressure a bad idea. We want to keep peripheral blood pressure at a low level, pull the pressure off the heart, delay or maybe totally prevent congestive heart failure. Okay, so she's involved in a whole diagnostic regime with a canine cardiologist thinking about this. The canine cardiologist proposes a drug to me—that white gel capsule on the table in front of you. She says, "Here is what Cayenne should be taking now, and you can experiment and find

*Companions in Conversation*

257

the lowest dose that works and go back up if you need to. And it's pretty effective. It's no longer made by the big pharmaceutical companies that used to make it; you have to get it from this small compounding pharmacy down the road here." And you feel like you're all of a sudden back in the early twentieth century, you know, before CVS, the small family compounding pharmacy, with a mortar and pestle in the back. But the reason they're doing this is that the high-priced molecules are no longer made by the big corporations if they're no longer profitable, okay?

But the drug in question, in this little white capsule, is a thing called diethylstilbestrol, or DES, and all of a sudden I had to hold my breath and hope she realized she was talking to a feminist of a certain generation, who had grown up in the women's health movement, within which prescribing of DES to pregnant women, supposedly to prevent miscarriages, had created terrible heritages of cancers in the adult children, reproductive malformations in both male and female children, on into the third generation. This was a *terrible* drug involved with a terrible scandal—big Pharma not releasing data, not responding to existing data, etc.—and it took a sustained women's health movement to expose it all. I also knew, because I'm interested in these things, that this drug had been developed in the animal industrial complex as a drug to promote weight gain in food animals. And on and on we go.

I knew a lot about these estrogen mimics, and both natural and synthetic estrogens, in both human *and* animal medicine and agriculture, and I was once again in the midst of extremely complex extractions of value from bodies—biopolitical, capitalist, Anthropocenic, whatever you want to call it. I was in the middle of the trouble with this little white gel capsule from my local family-owned compounding pharmacy that also sells homeopathic remedies. *And* DES is carcinogenic in dogs, too, albeit probably not in the doses and time schedules contemplated for Cayenne.

So you're always doing a balancing act with drugs. I said, Okay, I'll try this drug, and it turns out that all that is needed is a *very* low dose. But giving the dog of my heart this pill landed me in needing to write a DES manifesto, to go along with my own history with Premarin and pregnant mare's urine and all the cruelty and culpability in that terrible story. Cayenne and I were bonded in all the woes of female mammals, but for us in the historically situated land of biomedicine and biopolitics. I called the paper "Awash in Urine." It brought me back to the strong presence of American Jewish women in the women's health movement. It brought me back to the whole history of *which* women were brave enough to speak up, to make cross-gender, cross-race, and cross-species alliances, and to the racial/ethnic differences in feminism of that period. This little pill—my dog eating that little pill—brought us back into biopolitics with a vengeance.

cw: That reminds me of what I think is one of the smartest things that Roberto Esposito has said about biopolitics, which is that it

*Companions in Conversation*

doesn't operate at the level of "the person." It doesn't operate even at the level of "the body"—it operates at the level of what he calls "flesh."

DH: Yes.

CW: The level of what he calls "being-in-common." For biopolitics— you were talking about the extraction of value—species distinctions are not constitutive.

DH: They are not constitutive, they are *used*.

CW: They are used, and that extraction of value then ramifies differently for different people of different genders, of different races, of different species. But species itself is not the driver.

DH: No, it isn't the driver, nor is race the driver. I think "flesh" does something else, including making the shared tissues of race and species patent.

CW: But this makes you wonder why it took so long for people to realize that, with all the discourse about race, and race being so central to everything we've been talking about, you can't talk about race without talking about species.

DH: No, you certainly cannot.

CW: As I've often said, it's not for nothing our scholarship is called the "Humanities." It's amazing how long it took us to realize what is actually just a straight logical extension; I mean, we're not talking about a fancy two-step to get from race to species, and vice versa.

*Companions in Conversation*

DH: As you know I've never been happy with the term *posthumanism*. *Posthuman* we both find absurd.

cw: Right.

DH: But I've never rested easy with the term *posthumanism* either; I'm in alliance and disalliance.

cw: Yes, sure.

DH: I love your book, I love your analysis, and I understand how necessary posthumanism is, and practically all my friends are doing creative and necessary thinking under that sign, but I just can't. It was Rusten who said, "Well, it's not posthumanism, it's compost!" *(Both laughing.)* If you're in need of a slogan, "It's Not Posthumanism, It's Compost!" "It's Making Hot Compost! Compost Is Hot!" (which is Beth Stephens and Annie Sprinkle's phrase). Which then brings—and this is again Rusten—he says, "It's not humanities, it's humusities. It's *humus*." (cw *laughs*.)

Etymologically, the human is rooted in *humus*. Too many tones of "human" go to *homo*—which is the "bad" direction—but then there's "human" that goes to *humus,* which is the "good" direction. Not to be too simplistic about it. *(Both laugh.)* There's being part of the making of the soil and the earth and the *humus* direction, and there's the phallic "man" in the *homo* direction. (cw *laughing.*) There's the ever-parabolic tumescence and detumescence of *homo* in *that* direction of "humanities," but there are other possibilities in the humusities. So my slogan becomes "Not Posthumanist But Compost" *(both laughing).* I'm implicated in posthumanities, too, of course; I published *When Species Meet* under that sign, after all!

*(Many deep breaths and some well-aged Scotch later . . .)*

FOLLOWING FEMINIST THREADS

CW: Earlier, we left off talking about how the two manifestos ramify differently as individual documents, but how they also have pretty deep and unexpected connections that we tried to draw out, that bear on much of the contemporary interest in biopolitical thought, and even a little more broadly, biophilosophy and ecological thought.

DH: And bio-techno-political thought. The cyborg keeps making me remember the necessity of including—without assuming collapse into each other—the organic, the technical, the human and nonhuman, the many sorts of things that just don't resolve into binaries and are absolutely in what Marilyn Strathern might call relations of partial connection.

CW: Right. And in my mind there are really two main strands of biopolitical thought that we touched on earlier: one of a sort of Agambenian flavor, which is much more resolutely interested in—

in a very Heideggerian style—ontotheological questions and interested in the issue of sovereignty. And then there's the other strand
that involves you and Foucault and also what I was trying to do in
*Before the Law* in using some of the work in systems theory, including people like Luhmann, to actually extend and radicalize Foucault's work.

DH: Which is a strand I feel much more connected to than the
Agambenian line. I find some thoughtful threads in the Agambenian arguments; I also find various problems. But they don't
matter a lot to me; they're not really my problem, if you will.
The kinds of things you work on and the braids that you are doing are not the ones I'm doing, but I am deeply involved in them.
And there's a third line of the biopolitical for me, which really
does come through feminist lineages at their deepest. Some in
the academy, some not. Some deeply involved in questions in
ecological feminism, some located in questions of health, some
tied to questions of race, some in questions of our relations to
other sentient critters, the animal worlds, also plant, microbial
and fungal worlds.

There are many threads that themselves aren't the same, but
from early on I think my thickest thread—though I've been very
much involved in these other literatures and these other discussions—the ones that infuse all my thinking are first of all biological, including the systems thinking that comes through Lynn
Margulis.

cw: *Bio* bio-...

DH: Lynn Margulis, G. Evelyn Hutchinson, the biologist that is Gregory Bateson (that part of Gregory Bateson)—the systems theory that is important to me comes to me through that work, and then through the Macy conferences—Evelyn Hutchinson was a participant at one point—and the frog's eye/frog's brain work, and so forth. All of that really was where I came to systems theory, through biology. And then the work of Barbara Noske and Val Plumwood and other ecological feminist thinking, and the work of people like Deborah Bird Rose. You know, my lineages are truly, deeply feminist. And my citation of other feminist writers—mostly women but not all—is not a politically correct move; it really is where my thinking comes from. These are the people I think with.

cw: Right. And that's one of the reasons that another crucial resource, coming out of feminism, and going way way back, as you know, has to do with not being afraid of what's now being called an "affirmative" biopolitics, an affirmative sense of mortal connection with other forms of life. And then eventually, beyond that, realizing that within itself the feminism lineage needed some pretty radical queering to draw it away from biological ideas of reproduction and so on.

DH: There were also other strands within feminism. But yes to the need to interrupt the hypercritical, hyperventilating with critique.

*Companions in Conversation*

CW: I think that's why as a thinker you can be open to cross-species relations in a way that a thinker like Foucault never could be.

DH: Absolutely. Because another part of that was from the get-go an unembarrassed thinking with and through the love of nature: there is an affirmative feminist biopolitics, an affirmative relationship to worlding, to visions of the world and inhabiting the world that needs other critters. There is root feminist thinking here that is pretty well read out of the citational apparatus of most of the academic discussions.

## CATHOLIC FEMINISM, CATHOLIC SEMIOTICS: THE NEGATIVE WAY, IN THE FLESH

CW: On this note, I wanted to circle back to what in fact could be construed as a kind of Agambenian aspect of your work—although obviously a very complicated one (and not Heideggerian in the way that Agamben's certainly is). And that has to do with something that you don't shy away from talking about at all in the "Companion Species Manifesto" (quite the contrary): your Catholic background and your ongoing negotiation and navigation of Catholic thinking and what it's made available to you. I'm actually interested in *two* aspects of that, but the first—and this is the most conspicuous instance—has to do with what you call in the "Companion Species Manifesto" the "negative way of naming" or the "negative way of knowledge," which you also sometimes call a form of love.

*Companions in Conversation*

The second aspect—which you talk more about and is probably easier for readers to grasp—is this repeated motif of "the word made flesh." I want to hear you talk about this a bit, because "the word made flesh" pulls us in an opposite direction from "the negative way of thinking," as it's understood in negative theology. How does this relate you in a different way to thinking about biophilosophy, to thinking about life?

DH: And worlding within the finite, especially SF worlding—speculative fabulation, science fact, speculative feminism, science fiction.

CW: Yes, but how do those two strands relate and disrelate against the background of the bigger question, this Catholic thing?

DH: Well, those are two key strands. You know, folks who grew up Catholic and took it seriously the way I did went to Catholic schools, were deeply involved in it. I was really a believer and a practitioner for many years past adolescence. It was never a kind of trivial "Oh yeah, I went to Catholic school, and it really didn't mean much." Some folks experienced it that way, but that's not how I experienced it, at all. It really shaped me, profoundly. And some people who fall away from that . . . . First, the idea of "falling" away, are called "lapsed" or "fallen" Catholics, right? There's a whole joking culture about that, and Catholics can tell jokes about Catholicism they won't let anyone else tell. My hatred for the Church is probably much fiercer than that of people who were never in love with it.

*Companions in Conversation*

That said, I stopped calling myself and letting anyone else use the term *lapsed,* and I rather like the locution of *secular Catholic,* partly because of Susan Harding's influence and her insistence on the extraordinary importance of various modes of Protestantism in the formation of the American state, plus the situated co-constitution of religion and the secular. She is writing very provocative stuff right now about the Protestant secularists of both the Evangelical sort and the Enlightenment and scientist sorts, the extreme importance of Protestant formations to both of these—one acknowledged, one not. She tracks the tug between the contemporary Evangelical Protestant secularists, the secular Protestants, and the secular separatists. The secular separatists are those who really want to put into law and enforce the separation of church and state at every imaginable opportunity, schools being the main battleground. Anyway, Susan has really influenced my thinking.

I think of myself as a secular Catholic, *not* a part of the secular separatist Protestant scene, even if willy-nilly I have little choice about participating as an American. I am not a Richard Dawkins type of character, waging war against "irrationalism" (of course, he's British, so it's a little odd to bring him into the American context). But I approach religion out of quite a different formation.

DH: First of all, there's the material semiotics of Catholicism, which is the word-made-flesh part. One could say, for a moment, that the secular secularists—especially the secular Protestants of the European streams—developed the semiotics of the separation of the signifier and the signified, of the absolutely arbitrary relationship of the signifier and the signified, of the inability of the word to touch the flesh. This is the profound break with the sacramentalism of Catholic theology, which I think is embedded in what became the dominant mode of semiotics in the American university, quite different from Charles Peirce's semiotics, it must be said. (Remember, it is Peirce's semiotics and aspects of American pragmatism that shaped both Bruno Latour and Isabelle Stengers, both my close colleague-friends.) I think the traditions embedded in American pragmatism, although I don't imagine that their inspirations are Catholic, are compatible and were helpful for me. But the implosion of metaphor (and more than metaphor), of trope and world, the extraordinary tentacular closeness of processes of semiosis and fleshliness, sets me up at the level of both affect and cognitive apparatus for being suspicious of the division between the human and everybody else. And the division between mind and body within the human. It just sets me up for being really unhappy with those splittings and great divides, at a level of my most fundamental formation as a person in the world.

There's no question that explicit Catholic practice and intimate experience mattered. The powerful experience of first eating Jesus when I was seven years old—terrifying, wonderful, amazing. It was a practice and an experience of a very deep kind, at levels of visually vivid nightmares, fierce daylight plans, intense loves, relentless questions. And there is no question—again at the level of both affect and cognitive apparatus, the various purifications and sortings of the world, two by two, you know, nature/culture, biology/society, mind/body, animal/human, signifier/signified, nyeh-nyeh/nyeh-nyeh—I just really was never any good at all of that. That has deeply influenced who I am as a writer.

cw: Yes, and in one of the first things I wrote in so-called Animal Studies—the piece on *The Silence of the Lambs* that I wrote with Jonathan Elmer—I tried to zero in on how that is connected precisely to questions of flesh and of animality and of species through the figure of Hannibal Lecter. I mean, you talk about eating Jesus . . .

DH: Oh my.

cw: Lecter is a laboratory for all of the Enlightenment discourses that would, in bad faith—and he precisely outs the bad faith—

DH: Doesn't he ever.

cw: . . . of all these modes of separation and clean conscience. As so—

DH: That's a perfect example.

*Companions in Conversation*

CW: As we put in the essay, Lecter's position is not "I eat animals and *not* therefore humans"—he does not believe in that sacrificial substitution. His logic is, "I eat animals and therefore humans." So monstrosity ...

DH: Oh boy ...

CW: ... is always nearby when you're talking about—

DH: And there's no question that there's something about eating that outs this stuff particularly powerfully. "The Companion Species Manifesto" is deliberately working off of oral tropes: the first kissing scene, and the root meanings of companion, *cum panis,* with bread, at table together. I am deliberately working with questions of ingestion, digestion, indigestion.

CW: The importance of "messmates."

DH: Gestation, or *gestión,* bearing and carrying on, now not from the point of view of the uterus, but gestation from the point of view of eating, in substance this consubstantiality of eating and being eaten, which is different from the consubstantiality of either reproduction or generation. It just is.

CW: Yes.

DH: For me, the incarnation and sacramentalism were overwhelmingly about a shared meal, in and of the flesh. Carnality is seriously Catholic. Both cyborgs and dogs, both manifestos, bear witness to that!

cw: You know, in a way, "conventional" Christianity ends up being kind of piggy-in-the-middle on this—because, actually, Derrida ends up where you end up, but from the other end of the equation.

DH: He's coming at it as a Jewish Algerian, remember.

cw: So he's the crazy Jew and you're the crazy Catholic! (Laughing.)

DH: And Derrida had raised the question of sacrifice *way* more radically than I ever could have done it—I couldn't have even imagined the depth with which he got the structure of sacrifice, right? Derrida taught me—it's Derrida who led me to pose the problem as not "Thou shalt not kill" but "Thou shalt not make killable." We somehow must come to terms with these questions without the structure of sacrifice. It's not like we can just not inherit the structure of sacrifice—you don't have a choice of just setting these things down; you can't just set your burden down. But the question of sacrifice: for example, in science, the killing of the animal in the laboratory is still called "sacrificing." The only reason in industrial meat agriculture it's not called "sacrifice" is that the language is slightly secularized, but it's still practiced and perceived that way, it must be said. I think that Derrida's understanding and ability to communicate the depth of the trouble around the structures of sacrifice, beginning with Isaac, or before, and marching right through the Eucharist . . . (laughs)

*Companions in Conversation*

cw: We'd have to come up with a new name for you, because if Derrida, then, is "bad" (read: Algerian; read: Muslim, Jew), then what would the equivalent on the Catholic side be? *(Laughing.)*

DH: *(Laughs.)* Well, I'm not sure, but I know where I went—and this will take us back to Isabelle Stengers in a minute. But insofar as I would acknowledge a—um, oh, I don't know—the word *religion* is a truly weird one here.

cw: Yes.

DH: Because the whole category of religion is invented as a modernist category.

cw: Yes, it is so foreshortened.

DH: It's a modernist category in the same way that both "science" and "culture" also are. And we know its history. That said, where I am most, I don't know, at home—I am drawn toward the Earth religions, partly toward the Wiccan, and, with Stengers, toward Starhawk's practices, but especially toward the great, old and new, ongoing chthonic ones under and of the Earth. For me, it's not Marija Gimbutas's Great Mother. I've never been much attracted to the Great Mother worlds, although I think they are fascinating, the many stories about the invention of patriarchy on the destroyed body of mother goddess, and so on.

That said, when I say I am a creature of the mud not the sky,

I mean I am an entity given to the powers of Earth. I am terran. I am not astralized, not in awe of the chief gods and single gods, I am a terran. In league with the entities of Terra—Gaia is one of them, but Gaia is a bit of a problem if you go from Hesiod on, the *Theogony* on. By the way, I reread passages of the *Theogony* when I was thinking through some of these questions, and the hairs on my arms rose from the beauty of the language. And I was only reading it in a modern English translation, right? The power . . . I was just *stunned* by the beauty of the language. Wow. That said, Hesiod gave Gaia a cleaned-up lineage to ground the Greek pantheon and the Olympiad. Gaia is, shall we say, heteronormalized. That's not quite fair *(laughing)*. She's still plenty queer *(still laughing)*. But there is a kind of a heteronormative quality of the post-Hesiod Gaia that is hard to swallow.

cw: A little domesticated.

DH: Too much tamed. And I am much more interested in the lineages, or better webs, of Gaia that are not funneled into what becomes the Olympiad and the Greeks and the Romans and their Europeans. I am really interested in an older, wilder Gaia, in the Gorgones, the Nagas, Pachamama, Oya—in more and other than Gaia. Eduardo Viveiros de Castro and Déborah Danowski organized a meeting in Brazil around *Os Mil Nomes de Gaia*/The Thousand Names of Gaia—terran, global, heteroglot, finally unnamable. These names don't necessarily evoke or in-

*Companions in Conversation*

fluence each other; they may or may not be in historical, political, and cultural context; some are, some aren't. But I want to cast my lot with the ongoing, unfinished, dreadful powers of the Earth, where the risk, terror, and promise of uncategorizable mortal ongoing can still be found, and my Catholicism in the end went *there*. And I think that's kind of a natural turn—it's a naturalistic turn, among other things. The dreadful chthonic ones aren't transcendent, they aren't gods, they aren't omniscient beings, they aren't fixed entities, they aren't objects, they don't call for religions, much less beliefs. These are names of powers—or maybe the unnaming of fixed powers—and this gets me to the negative way of naming.

cw: Well, this is what I was going to come back to, because if all of this takes you back toward the mud, and back toward the dirt . . .

DH: . . . with the pigs and their peoples and carnalities, the ones the Monotheists truly couldn't handle *(laughing)* . . .

cw: . . . I mean, I think to talk about the word made flesh and to talk about Catholicism is first and foremost to talk about the word as fetish. If I think back over the length of your career, you have coined a lot of very powerful terms.

DH: I prefer them to words that granulate in your hands.

cw: But a lot of these terms have a kind of fetishistic power, I think, for a lot of people.

DH: Knowing and not knowing the collected knowledges, the investments of desire . . .

CW: Right, and there are a zillion of these terms.

DH: And as soon as you fix them, or singularize them, then you're involved in idolatry and, in a way, fetishism. Besides, you never have a correct love, because love is always inappropriate, never proper, never clean—that's deep in my writing.

CW: Yes, and so I think the word made flesh is a way of marking a relation to doing, let's say, biophilosophy and technofeminism, queer biophilosophy, as a writing practice.

DH: I think that's true.

CW: But here's the interesting thing—

DH: But before you go to that, the word made flesh: you know, John's is not one of the synoptic gospels. The "word made flesh" is a very problematic phrase.

CW: Yes. Yes, it is.

DH: And so, I know that. And so I'm using it with a kind of . . . I'm using it and not using it at the same time, again tied into the fetishism question. . . . But you were going somewhere else.

*Companions in Conversation*

cw: Well, what I was circling back to is that then one has to ask—and this I think is really fascinating—one has to ask, What is this "other" (so to speak) commitment, which is to a structure familiar to us from negative theology, of the negative way of knowing? And you actually use the term *theological*.

DH: Absolutely, I read the theological works in question as a kid in college, just absolutely enamored.

cw: Right, but to me—and maybe you're coming at this from a different direction—but to me, the negative way of naming . . .

DH: . . . is in a generative friction with . . .

cw: . . or it doesn't necessarily direct you away from Earth, but it directs you toward a kind of knowledge that can never be made manifest in flesh. Are you using it that way, or differently? What's underneath that?

DH: A little differently from what you just said. I understand the suspicion or just flat-out contradiction that you're asking me about.

cw: Oh, I don't think it's a contradiction, and I can tell you why in a minute, but go ahead.

DH: Here's how I think about it. It's not that the word is made *manifest* in flesh. It's that semiosis and flesh are—what?—not one, not two . . . what can we say next? It's not that something is made manifest in something else at some level deeper than symbol. Well, what's that? There's some more radical structure of identity/nonidentity; there's some more radical structure of nonidentity here that is profoundly materialist. And that's a problem for names. The minute you name something like that, you have misnamed it. The minute I name the chthonic powers, I have, by the very name itself, committed a kind of fixing of a fetish, a kind of idolatry. I think that this is where the *ouroboros* swallows its tail. I think that by going into the mud, into this proliferation of words—I think my proliferating words and figures themselves are flesh and do a lot of things. But what they can't do is stay still as a conceptual apparatus that makes most philosophers happy, and so they end up saying, "*Mere* metaphor," and I think, "Give me a break, guys. This is not mere metaphor, this is actually an enactment of, among other things, corporeal cognitive practice."

CW: Right, well, this is why I mentioned fetish earlier, because the first thing about the fetish is that it materializes something that is beyond the site of materialization. But it's not—and this comes back to the "negative way"—it's not a given beyond, and it's not a fixed beyond, and it's not an antecedent beyond.

DH: Well, the negative way is a mode of thought that was originally developed in relation to the question of God.

CW: Of course.

## Companions in Conversation

DH: The definition of God . . . any effort to produce a positive theology of God fails from the get-go, because God exceeds all possible specification; all possible names are from the get-go defeated because of the exeedingness of that which cannot be named. And you can't have an object without a proposition, so you call it that which cannot be named, which is already wrong. Even *that's* wrong. The infinitude of this, the infinitude of the nonpositivity.

cw: Oh no, that makes perfect sense to me.

DH: Well, it makes perfect sense to me, too, but I assure you that we're in the minority! *(Laughter.)* Well, transpose that, when the problem isn't any more God or Being, or infinity, it's actually finitude and mortality. The negative way of naming in theology was developed around the problem of infinity. I think for me the problem is, well, the binary opposite—you know, which is sort of embarrassing to say because, well, you can readily see why *(both laughing)*. I mean, you laugh when this happens to you; language does this to you.

cw: Of course! But you know, when you were talking earlier about your interest (and this is before you were talking about Gaia) . . .

DH: . . . the chthonic ones . . .

cw: Yes, but I think the way that I see the negative way, negative naming, functioning in your work is that it's going to insist on a distinction between what you're doing and the idea of some kind of finally holistic Mother Earth.

DH: Absolutely. You will not come together from two, or many, into one, because that is precisely the idolatry that the negative way tries to block.

CW: That's right. And so, to me, here we actually circle back to what seems a very deep connection between your work and later-generation systems theory of the kind that Niklas Luhmann is doing, because Luhmann once said that the closest thing to the second-order systems theory he does is the negative theology of Nicholas of Cusa.

DH: Well, okay.

CW: And I actually think that, for the very same reason, what you're calling in your work the word made flesh is a kind of materialization of something that's also radically not present because it's bigger. But it's not bigger in the sense of "Oh, you can point to it and grasp it."

DH: It cannot be dealt with indexically, it cannot be dealt with holistically, it cannot be dealt with representationally. I mean, I think the negative way is a terribly serious injunction to, among other things, humility.

CW: Yes. Yes.

DH: It was like that question at the conference over the weekend. What do you do when your tools hit the wall? The negative way is constantly asking the question, What do you do when your tools hit the wall? When I say that I'm a creature of the mud, I am *of* the mud—forget the word *creature*—I am *of* the mud, the muddiness is ongoing. The worlding, the sympoiesis . . .

## Companions in Conversation

cw: ... the muddling ...

DH: ... truly, I am muddling, and I am in the muddle (I used that in a lecture title recently; *muddle* is a fascinating word). So "muddling along" is taken as the definition of not thinking, when it's quite other than that. So I think we're in a *ouroboros*, snake-swallowing-its-own-tail kind of moment with this commitment to semiotic fleshliness, which I am saying instead of using John's "word made flesh" because I want to get away a little bit from the particular track that took, theologically. The semiotic fleshliness, what I ended up calling "the material semiotic," the semiotic material, the inextricability of it.

cw: You want to get away from the done deal theologically.

DH: Well, I want to get away from the Hellenism through which John comes down to us. I want to get away from that particular theological tradition. And in the mud, or in the muddle, full of tentacular ones, including *ouroboros,* the snake is always swallowing its own tail. That *can* be taken as a figure of a great completion.

cw: As a figure of holism, yes.

DH: But it shouldn't be. Among other things, the snake—well, we're going to have problems of excretion in the end because there's another hole! I mean, we've got another hole operating here! As soon as you take the snake seriously, then you can't use

that as a figure of holism. But you *can* take it as a figure of a certain way that the material semiotic, the fleshly semiosis, meets
the negative way. There is a kind of *ouroboric* quality, keeping in
mind that you can't have the figure of the whole Earth, whether
it's whole Earth of NASA, and of a certain kind of (misinter-
preted, Latour argues persuasively) Lovelockean Gaia hypothesis about the living Earth. You cannot have the whole Earth
either way, either from the older traditions or from the more
space-age formulations.

CW: And that's precisely why you can neither be utopian nor dys-
topian.

DH: No.

CW: Because what's at stake—this is, to me, a very strong point of
contact between what we've been talking about in your work and
the kind of stuff Derrida does—what's at stake is futurity and *making*
futurity.

DH: I agree with that.

CW: Precisely because it's not about infinitude in some way.

DH: It's not about past-present-future.

CW: No. And so that brings the emphasis back to this kind of dynamic
process of materialization.

DH: I agree with that, and I love Derrida in just this way. And you're one of the people who give me Derrida, who make me need and want to read what I would probably otherwise just say, "I already know this. There. I'm sure it's great." *(Both laughing.)* But what I really *do* read . . .

cw: . . . but I'm reading about cephalopods right now!

DH: . . . *I'm* reading about cephalopods right now, goddamn it! *(Both laughing.)* And not only that, I am reading Ursula LeGuin again, and I get so much from her! Truly, with every bit as much nuance and depth. She and I just had this little email riff together today around the storying and caring of Earthlings, and she wanted to put "music-ing" in there. And so on. Without dystopia *or* utopia. Her kind of "always coming home." And remember *The Word for World Is Forest,* where at the end of this book—that the blockbuster film *Avatar* did such violence with and didn't even deal with the intellectual property rights around—anyway, that's another issue around LeGuin's story. But at the end of *The Word for World Is Forest,* the indigenous leader says that we can no longer pretend that we don't know how to murder *each other.* LeGuin, like Derrida, cannot rest in, cannot have the solace of, a utopic future. I turn to what Deborah Bird Rose would call with her Australian Aboriginal teachers and interlocutors, somehow being response-able in the thick present, so as to leave more quiet country to those who come after; you're facing those who came before. Anyway, I've learned from many writers that resonate with Derrida. *And* I have high stakes in citation apparatuses, since Derrida gets cited *a lot,* as a theorist, and Ursula le Guin, kinda never. I have straight-up old-fashioned feminist stakes in citing accurately where I get my ideas. In brotherly love with Derrida, but not from him *(laughing).* My sisters rock!

CW: Although he would probably accept her income on her novels, I would guess *(laughing).*

DH: Well, but you see where I'm at; I'm joking about this a bit, but I had an elite education, too. I was reading the medieval theologians, I was reading Heidegger, I was reading Jaspers, I was reading biology, and James Joyce. I mean I have a perfectly elite education, thank you, thanks to Sputnik. My Catholic girl's brain got educated, as opposed to my being a pro–Life activist mother of ten, because I became a national resource after Sputnik. My brain got valuable, and so I got this crazy education instead of being an Irish Catholic pro–Life activist.

CW: So you're a Sputnik Catholic!

DH: I'm a Sputnik Catholic! I mean, there was a branch point, and that branch point wasn't about me being a neat person or something. It was about—I became a national resource, at a certain moment in the Cold War *(both laughing).* Very humbling!

CW: I think that's a great way to frame it, and I think this a great aspect of your work, and we could actually say more—and maybe you do want to say more—about that. I think that irrigating and aerating this force of the Catholic in your work is something we've done with your work, but, you know, you're not the only one, as you've pointed out. I mean, there's Isabelle Stengers, there's Latour.

DH: Hey, the Catholic thing turned out to be kind of big!

CW: It makes me want to say, "Hey, what's going on here?!" *(Laughing.)*

DH: Whoa! *(Laughing.)* Wait a minute, the Jews and the Catholics are truly taking over! I think there's something to that. I think the Protestants ran out of steam, thank God, and none too soon. I'm kidding, of course.

CW: But it is kind of an interesting phenomenon.

DH: But we haven't even begun to talk about, we have not touched the extraordinary calling to account from none of the above, thank you—the thinkers of various indigenous traditions who are also in the written record, and we can no longer ignore them. The folks out of these traditions are themselves reworking their current and past heritages, not to mention thinkers from globally diverse Islam and . . .

CW: . . . that would be another manifesto . . .

DH: This is a really big deal: we have not even begun to talk about the other great literate traditions, thank you very much. We are being very parochial and we are acknowledging it, right off the top.

*Companions in Conversation*

cw: Yes, we are.

DH: That said, the Catholics and the Jews are taking over! *(Both laughing.)* At last, and thank God I'm here at the time! *(Laughing continues.)*

## COSMOPOLITICS, COMPOSITION, COMPOST

cw: Well, I wanted to maybe wrap up by asking . . . . We mentioned Isabelle. As I've told you, one of my favorite moments in the *Cosmopolitics* project that we did in the *Posthumanities* series in translation is her engagement of Richard Dawkins's attack, essentially, on religion and the kind of cosmopolitical response that she offers to that, which I think is really—

DH: It's so fundamental.

cw: It's really remarkably thoughtful, very nimble, but I also think it's very powerful and very pointed in places. And I wanted to ask you—we've been talking about philosophy, we've been talking about theory, we've been talking about issues that involve certain kinds of audiences with certain kinds of expertise and not others, which is a real issue, as we have learned from your work and from lots of people, in terms of political effectivity and making social change. I wanted to ask you a question that I really don't have an answer for. What would a cosmopolitcal response look like to the fact that in the United States at the moment, apparently 50 percent

*Companions in Conversation*

or more of the people in the country believe in Creationism? And probably those same 50 percent or more say that global warming doesn't exist. What is the cosmopolitcal response to the situation in which half the country believes this and thinks that the other half is crazy and vice versa?

DH: And vice versa.

CW: I think this is a huge question.

DH: No, I couldn't agree more. I think it's urgent, and it's only one of a deck of cards of questions that are linked in this kind of structure, so . . .

CW: . . . so where do we begin?

DH: I think that there are some places to start, and I think there are some people who have started, and with whom we must connect and enlarge and think. And I'd start with Isabelle's "not so fast." The pluralist imagination has always imagined that if you could just get people to sit down at the same table together and they could just talk to each other for long enough, they would somehow come to understand each other well enough that they could make decisions in the common good. That's the fundamental democratic liberal pluralist model, which is clearly broken, and, you know, who could not have a soft spot in their heart for that model? We know its problems, but losing it is not a small problem.

So Isabelle, in her cosmopolitical thinking, makes us pay attention to this: what about the folks who really want to say, "Not me, thank you, not your table, count me out. You may think you're endlessly inclusive, but frankly, count me out, and not so fast." I think that's a little bit the structure of what we're looking at here. There are many sides of this fundamental split. There are misidentifications of very important kinds. For example, on the global warming/climate change complex. Susan Harding is one of the people who insists that a lot of folks who say this is a conspiracy, or that this isn't happening, or maybe that "God wouldn't let that happen" and Providence will provide, or they're "science deniers" or whatever—a lot of these folks, what they're really mad at is Big Government, and a particular sense that Big Government has always screwed us over, and Big Science has always screwed us over. Of course, these same people may be accepting some sort of major agricultural subsidies. Think how subsidized the economy of Kansas is, for example, with federal dollars. But Susan says that all this gets ascribed to a fundamentally religious thing when a lot is going on, and it matters to be precise about what's happening. And then, within these matters it is important to notice—let's take Kansas again—the "creation/care" people, who are really upset at the failure of the ethical obligation of stewardship as Christians and are working very hard on such things as the better care of animals and not screwing up the climate. Raise "evolution" and they're out the door, but raise questions of good stewardship and you've got a practical conversation going on.

*Companions in Conversation*

So why raise all the questions at once? Why not be willing to disaggregate what you're so sure of? Me, I'm a scientist—I'm really very sure about the evolutionary history of life on Earth. And I'm really very sure that the climate modelers are more right than wrong. You know, I'm really pretty sure of a lot of things, because I think, in Bruno's sense, the networks are very strong. This stuff holds against strong tests. Well, people like me, which is half the country, need to be willing to disaggregate a bit and engage what Isabelle in her cosmopolitical thinking will call an "ecology of practices." Okay, here we are in the Central Coast of California in a big drought. Let's think about water. Let's not think about water by saying in your first sentence, you know, "caused by global warming." Some people are going to think that, and some people, not. . . . What we're worried about together in our communities is water. That's already hard enough.

CW: Let's start with a problem that we all agree we share.

DH: We all share this problem, and we all have very different ideas about what to do about it. That's already hard enough. That does not mean the science is not settled on climate change, or that relativism reigns; it does mean learning to compose possible ongoingness inside relentlessly diffracting worlds. And we need resolutely to keep cosmopolitical practices going here, focusing on those practices that can build a common-enough world. Bruno says this, too. Common is not capital *C* "Com-

*Companions in Conversation*

mon." How can we build—compose—a better water policy in the state of California and its various, many parts? How can we truly learn to compose rather than decry or impose?

CW: And I have to ask—and we've talked about this over the past couple of days in terms of the very important term "we"—who's the "we" here (which to me is a term of audience)?

DH: And there are people who put their body on the line and say, "I don't want to be part of this process."

CW: That's right. And also it's in turn a question, often, not of theory but of rhetoric. And if you don't pay attention to audience and to rhetoric—and I'm speaking now partly based on my experience as an animal rights activist twenty, twenty-five years ago—one thing you learn very quickly is that if you can't use a different rhetorical toolbox with different audiences—

DH: You're not very good at what you do . . .

CW: . . . then you're never going to get anywhere.

DH: Well, if you can't use a different rhetorical approach and get a different toolbox, then you don't care very much about the animals. And perhaps also, a different *ontological* approach, attuned to different compositions, different worldings.

CW: That's right. And this is where I find Isabelle's work more useful than Bruno's.

DH: I think Isabelle's thinking is very radical.

cw: I think she has a better ear—not just a better ear but maybe also what you could call a better sensibility for just how nimble and supple and, as you say, sort of hesitating and self-questioning these rhetorical issues are.

DH: There's a huge overlap between Isabelle and Bruno in this, too.

cw: Oh, of course, huge.

DH: They have been in deep, thick, loving exchange for years.

cw: Of course.

DH: That said . . .

cw: They do different kinds of work.

DH: They do different kinds of work, and also they draw from different communities of practice in their thinking—and this is where I think Isabelle is drawing from the work of Starhawk—not just Prigogine, Deleuze, Whitehead, and—

cw: I'll tell you who else she and I talked about the last time I saw her: William James.

DH: Absolutely. James is terribly important to her.

cw: And Pragmatism. I told Isabelle, "I think of you as, first and foremost, a pragmatist thinker."

DH: I'm sure she accepted that.

CW: She not only accepted it; she said, "This is a philosophical resource and tradition that we don't have in Europe that is really, really important to me, and I had to come to people like James, but more broadly the pragmatist tradition, to get there."

DH: Yes. So look: Deleuze, James, and the traditions of practice, she is interested in *practices*. So she's not interested in Wicca as religion: she's interested in the practices that gather up and make worlds. All three of those—Deleuze, James, and Starhawk—are illustrations of partial differences between her approach and Bruno's. I think they appreciate this in each other. There's this huge overlap that they share around questions of actor-network—well, semiotics. "Actor-network theory" is much too reductive.

CW: Semiotics in the most general sense.

DH: And semiotics in the Peircean tradition. Actually in that sense, Bruno and Isabelle converge around inheritances from pragmatism. I think Bruno has been resistant, for diverse situated reasons, to certain of the resources that are important for thinking in the present mix. Foucault is one of them; Marx is another. The entire feminist tradition has been another, but I see that changing now. Bruno has become much more aware of feminist thinking, and curious, but it's been very hard for him actually to *use* the work in his own arguments and figures. He cites the work more richly now, but actually *using* it is just beginning. But why? I don't entirely understand this history. I consider him a serious friend as well as interlocutor. At one time, he was seriously upset with me for what he called (or Jim Clifford actually called) the "kitchen-sink syndrome." Because I want everything, I end up putting it all in! *(Both laughing.)* But he's a more careful thinker! Perhaps in less of a muddle.

CW: I actually would not say that he's a more careful thinker, but I know what you mean.

DH: You know what I'm saying.

CW: I know what you mean, and my guess is that people would agree with you. I mean, for me, thought practices are writing practices— as you said, they're practices of materialization. If there's no other lesson in twentieth-century philosophy, that's the lesson.

DH: Yes.

CW: That's why Heidegger says things like "the world worlds," and that's why we get the entire lineage of thinking that we do.

DH: Absolutely.

CW: And that's why we get you treating what most people call "signifiers" as fetishes in "the word made flesh."

DH: Yes, I'm doing something different.

CW: So to me, you're just doing something different from Bruno.

DH: Tremendously interlaced, but also quite different.

CW: Theoretically interlaced, but actually as a practice quite different.

DH: As a practice, different work. But watch for a minute Bruno's work at the Sciences Po in Paris, and his AIME project, and Isabelle's work with GECO, the *groupe d'etudes constructivistes* in Brussels. Both of them have been really engaged—in the writing and also more than that. For example, in Bruno's case, in engagements in theater practice, projects with earth scientists, engagements in new ways of trying to pull together worlds that make a difference. There is a way in which Bruno's practice has been, in my view, very much in the right sort of muddle. And Isabelle's, too. And it shows up also in the writing of their collaborators, students, and associates. It shows up in their projects. I think that Bruno in Paris and Isabelle in Brussels have been important in nurturing many kinds of generative work. I feel like we are in a string figure game with each other.

CW: And you're holding down the Catholic fort in North America! *(Both laughing.)*

DH: Or at least in Santa Cruz!

*Companions in Conversation*

DH: I want to end our conversation with the seed of a "Chthulucene Manifesto." *My* Chthulucene is the time of mortal compositions at stake to and with each other. This epoch is the kainos(-cene) of the ongoing powers that are terra, of the myriad tentacular ones in all their diffracted, webbed temporalities, spacialties, and materialities. *Kainos* is the temporality of the thick, fibrous, and lumpy "now," which is ancient and not. The Chthulucene is a now that has been, is now, and is yet to come. The Chthulucene is a relentlessly diffracted time–space (remember Karen Barad on quantum fields). These powers surge through all that are terra. They are destructive/generative and in no one's back pocket. They are not finished, and they can be dreadful. Their resurgence can be dreadful. Hope is not their genre, but demanding response-abilities might be. Terran forces will kill fools who provoke without ceasing. Killed but not gone, these fools will haunt in tentacular ongoing destruction.

The chthonic powers, both generative and destructive, are kin to Latour's and Stengers's Gaia. They are not mother; they are snakey gorgons like the untamed and mortal Medusa; they do not care about the thing that calls itself the Anthropos, the upward-looking one. That upward-looking one has no idea how to go visiting, how to be polite, how to practice curiosity without sadism (remember Vinciane Despret and Hannah Arendt). In the Anthropocene (a naming I have come to need, too), the

*Companions in Conversation*

chthonic entities can and do join in accelerating double death provoked by the arrogance of the industrializers, supertrans-porters, and capitalizers—in seas, lands, airs, and waters. In the Anthropocene the tentacular ones are nuclear and carbon fire; they burn fossil-making man, who obsessively burns more and more fossils, making ever more fossils in a grim mockery of Earth's energies. In the Anthropocene, the chthonic ones are active, too; all the action is not human, to say the least. And written into the rocks and the chemistry of the seas, the surging powers are dreadful. Double death is in love with haunted voids.

The chthonic ones can and do infuse all of terra, including its human people, who become-with a vast motley of others. All of these beings live and die, and can live and die well, can flourish, not without pain and mortality, but without practicing double death for a living. Terran ones, including human people, can strengthen the resurgence (Anna Tsing's kind) of vitalities that feed the hungers of a diverse and luxuriating world. The Chthu-lucene was, is, and can still be full of "Holocene" resurgence—of the ongoingness—of a wild, cultivated and uncultivated, dangerous, but plentiful Earth for always evolving critters in-cluding human people. Mixed and dangerous, the Chthulucene is the temporality of our home world, terra. All of us who care about recuperation, partial connections, and resurgence must learn to live and die well in the entanglements of the tentacular without always seeking to cut and bind everything in our way. Tentacles are feelers; they are studded with stingers; they taste

*Companions in Conversation*

the world. Human people are in/of the holobiome of the tentacular, and the burning and extracting times of the Anthropos are like monocultural plantations and slime mats where once forests, farms, and coral reefs flourished, which were allied to fungal materialities and temporalities in very different ways.

The Anthropocene will be short. It is more a boundary event, like the K-Pg boundary (Cretaceous–Paleogene boundary), than an epoch. This is Scott Gilbert's suggestion. Another mutation of the thick *Kainos* is already coming. The only question is, Will the brevity of the Anthropocene/Capitalocene/Plantationocene "boundary event" be because double death reigns everywhere, even in the tombs of the Anthropos and his kin? Or because multispecies entities, including human people, allied in the nick of time with the generative powers of the Chthulucene, to power resurgence and partial healing in the face of irreversible loss, so that rich worldings of old and new kinds took root? Compost, not posthuman.

The Chthulucene is full of storytellers. Ursula LeGuin is one of the best, in everything she wrote. Hayao Miyazaki is another: remember *Nausicaä of the Valley of the Wind*. And then go to the Iñupiaq online game *Never Alone*. Watch the trailer! http://neveralonegame.com/

With these storytellers, my next manifesto is "Make Kin Not Babies!"

# ACKNOWLEDGMENTS

It is impossible even to begin to name and thank all the friends, students, colleagues, kin, comrades, critics, librarians, dogs, dog people, nondog critters, publishers, editors, SF writers, artists, social movements, bookstores, marketing people, universities, cultural institutions, laboratories, funders, translators, and both friendly and malignant digital entities that made *Manifestly Haraway* take shape. I tried in citations to hint at the extraordinary degree of reciprocal generation and collective composition evident in my work from the very beginning. These two manifestos are rooted in big, bumptious communities, and they were written to and for those communities. My gratitude is exceeded only by my debts. But citations do not do justice to the publishers and editors who populate every page through their intensely personal work practice. Here, I want to name a few of the extraordinarily generous and skilled people who have made my books and essays grow into public objects over four decades.

A comrade and colleague whom I miss terribly, Jeffrey Escoffier, solicited and edited "The Cyborg Manifesto" for *Socialist Review* in the early 1980s. Everything that followed should be blamed on him. William Germano nurtured my work from *Primate Visions* through *Modest_Witness@Second_Millennium.* Because he made so many real books at Routledge flourish for so many of us for so long, he shielded us from the neoliberal corporatization that now threatens serious publishing everywhere. Conducting and editing the book-length interview that became *How Like a Leaf,* Thyrza Nichols Goodeve showed me what collaboration in a published conversation can be. Marshall Sahlins, himself a man supported by Great Pyrenees guardian dogs, solicited and published *The Companion Species Manifesto* for Prickly Paradigm Press; he, too, along with Matthew Engelke, must be blamed for a great deal that followed. Cary Wolfe solicited and supported *When Species Meet,* and he spent many days planning, interviewing, conversing, and editing our joint text for *Manifestly Haraway.* (Cary and I would like to thank Seth Morton in the Rice University English department for transcribing the nearly five hours of conversation that eventuated in the last portion of this book.) Cary's friendship and colleagueship, also enabled by excellent dogs, are a rare gift. Not least, Cary brought me into Doug Armato's extraordinary care at the University of Minnesota Press. Doug, with close colleagues at other invaluable university presses, continues to make it possible for scholarly books to flourish. In particular, Ken Wissoker at Duke University Press has shepherded an extraordinary number of books in which my favorite authors and essays (and even many of my own) appear.

I have not begun to name all the people, from copy editors through junior and senior colleagues, who helped turn my excessive manifestos into readable text, page by page. Excesses remain, and that is not their fault. I hope that they all know that they are manifestly, excessively appreciated!

# INDEX

*Index*

*Index*

California: drought in, 288–89

California Gold Rush: Australian Shepherds and, 94, 172, 174–75

canine evolution, 117–24. *See also* dogs

Canine Genetics Discussion List, 131

caninophiliac narcissism: neurosis of, 124

capitalism: advanced, 69n5; in China, 243; as historical system phenomenon, 239–40; homework economy as world capitalist organizational structure, 38–45, 73n21; late, 31, 50, 53; three major stages of, forms of families relating to, 40–41; violence of making live for profit, 229, 231

Capitalocene, 238–43; players in, 240; temporal synchronicities in, 240–41

Carby, Hazel, 75n32

Cargill, John, 105

Catholic background, influence of, 265–69, 270, 271, 272, 274, 283–85, 293; Catholic feminism, 265–67; doctrine of the Real Presence, 107; "lapsed" or "fallen" Catholics, idea of, 266;

sacramentalism of theology, 268–69, 270; secular Catholic, Haraway as, 267; semiotics of Catholicism (word made flesh), xii–xiii, 107, 110, 266, 268–69, 274–75, 279, 280, 292; Sputnik's impact on U.S. national science education policy, 51, 283

Caudill, Susan, 130, 189

Cayenne Pepper, Ms. (dog), 93–95, 176, 178, 190–92, 221, 222, 224, 247; as elder dog, drugs for, 257–59; microchip injected under neck skin of, 93, 132, 222; training, 131–34, 148, 151

C3I (command-control communication-intelligence), 6, 34, 205, 219; feminist cyborg stories recoding communication and intelligence to subvert command and control, 56

Charnas, Suzy McKee, 74n28

Chicana constructions of identity: Malinche story and, 56–57

children: loving dogs as, 124, 127, 128–29

China: capitalism in, 243

Christian, Barbara, 75n32, 77

Christian Creationism, 10, 286–88

*Index*

305

*Index*

community: monsters defining limits of, 64–65

companion animals, 215; communication between humans and, 140–45; companion "animal happiness," Hearne on, 143–45; emergence of term, 104–5, 106, 244; human obligation to, 145; talent, 143

companion species: cyborgs within taxon of, 96, 103, 113–14; human–landscape couplings within category of, 114–16; kin genres of, need for other nouns and pronouns for, 187

"Companion Species Manifesto, The" (Haraway), x–xi, xii, 91–198; affirmative biopolitics and finitude, 226–28; biopolitical worldings, 220–23; breed stories, 154–93; contingent foundations, 98, 101, 103, 116; core of, 152–53; emergent naturecultures, 93–117, 221; evolution stories, 117–24, 154; love stories, 124–31, 154; making kin, 223–26, 252–56, 296; narrative voice in, 218–20; negative way of knowledge in, 141, 265; oral tropes in, 270; ordinary biopoli-

tics and, 244–48; pet medicine and right/obligation to health, 256–60; practice of joy in, 128, 129, 152, 153, 227, 244, 252–56; questions explored in, 95; sidewinding sympoiesis, 223–26; situating in context, 214–20; as story of biopower and biosociality, xi, 97, 106, 123, 182; training stories, 132–54, 154

composing possible ongoingness inside relentlessly diffracting worlds, 288–89

Computer Professionals for Social Responsibility, 74n26

Connery, Chris, 243

conquest: legacy of, 94, 157, 220, 223

consciousness: cyborgs and, 13; false, 26, 50; gender, race, or class, 16; implicated in globalization processes, recognizing emergent forms of, 155; MacKinnon's theory of, 24; of non-innocence of the category "woman," 21; oppositional, 17–19

consubstantiality of eating and being eaten, 270–71

consumption: impact of social

*Index*

307

*Index*

cyborg heteroglossia, 68, 70n5,
75n29
"Cyborg Manifesto, The" (Haraway), 3–89, 102; citational
practice in, ix; fractured identities in, 16–27; homework
economy "outside the home,"
37–45, 73n20–21; informatics of domination in, 28–37;
microchip injected under
Cayenne's neck skin as direct
thread to, 93, 222; myth of
political identity, 52–68;
original context of, 201–7; as
phenomenon, vii–viii; as product of its moment, ix–x; purpose of writing, 96; range of
tones, personae, and voices in,
viii–ix; reception of, 210–11;
refiguring irony in, 20–21, 52,
65, 96, 208–12, 221; "spiral
dance" ending, x, 14, 68;
women in integrated circuit,
36–37, 45–52; "word made
flesh" trope used in, xiii;
working toward unknowing
in, 212–14
cyborg politics, 57–58
cyborg world: defining contradictions of, 63–64; perspectives

on, political struggle to see
from both, 15–16
cyborg writing, 55–59; poetry,
70n5; about power to survive,
55–58, 62, 65

Danowski, Déborah, 273
Darwin, Charles, 106
Daston, Lorraine J., 76n36
Dawkins, Richard, 267, 285
Dawn (Butler), 63
death: double, 232, 295, 296; letting die, 234; making killable,
233, 234, 235–36, 271; sacrifice,
question of, 271
Death of Nature, The (Merchant), 14
de Bylandt, Conte Henri, 160
decolonization: polyphony
emerging from, 19
de Courtivron, Isabelle, 75n29
Defenders of Wildlife, 170
De Fontenay Kennel, 160
de la Cruz, Catherine, 161–64
Delany, Samuel R., 52, 62, 74n28
de Lauretis, Teresa, 72n15
Deleuze, Gilles, 291
democratic liberal pluralist model,
286
Derrida, Jacques, 75n31, 228, 229,

233, 234, 271–72, 281–82, 283; on our "life" predicated on violence of massive "letting die," 234; on question of sacrifice, 271; "writing as a Jew," 230

deskilling, 39

Despret, Vinciane, 294

destructive species: Anthropocene and, 237–43

de Waal, Frans, 12

diethylstilbestrol (DES), 258–59

differential reproduction, 121

Directory of the Network for the Ethnographic Study of Science, Technology, and Organization (1984), 69n4

*Discipline and Punish* (Foucault), 69n5

disciplined spontaneity: agility training and, 154

DNA studies of dogs, 120–21

dogs: adoption of, 185–88, 251–52; Australian Shepherds, 94, 155, 172–79; Border Collies of Scotland, 115–16, 131; breed stories, 154–93; category of one's own ("unregistered"), 155, 179–90; as first domestic animals, 119; foundation, 176; functions of, throughout history, 104–5; Great Pyrenees, 155, 156–71; herders, 155, 172–79; heterogeneous history of, in symbol and story, 105–6; importance of jobs to, 131; on the island of Borneo, 248; livestock guardian (LGDs), 155, 157–71; love stories about, 124–31, 154; Native American dogs, 105, 174, 189; Navajo dogs, 105, 165; relationship with human beings, 103; rescue, 184–86, 251–52; training stories about, 132–54, 154. *See also* "Companion Species Manifesto, The" (Haraway)

*Dogs* (Coppinger and Coppinger), 156, 166

Dog Working Trials, 148

domestication, 142; of dogs, 119, 120–21, 122; as emergent process of cohabiting, 122; as evolutionary strategy, 122

domination(s): organics of, 28–30, 32; of white capitalist patriarchy, 69n5. *See also* informatics of domination

D'Onofrio-Flores, Pamela, 73n19

double death, 232, 295, 296

*Double Helix Network News,* 177

*Index*

*Index*

Gates, Henry Louis, Jr., 75n31

GECO (*groupe d'etudes construc-tivistes* in Brussels), 293

gender: cyborg, 66; in feminist science fiction, 62; "New Industrial Revolution" and homework economy, 37–45, 73n20–21; sociobiological origin stories and dialectic of domination of male and female gender roles, 43, 74n23

gene flow, multidirectional, 101

genetic determinism, 123

*Genewatch, A Bulletin of the Committee for Responsible Genetics,* 73n18

genocides, 227; living in times of, 231–32

Giddings, Paula, 75n32

Gilbert, Sandra M., 75n32

Gilbert, Scott F., 123–24, 242, 250

Gillespie, Dair, 151

Gimbutas, Marija, 272

*Global Electronics Newsletter,* 68n4

globalization: recognizing emergent forms of consciousness implicated in, 155

global warming: cosmopolitical

response to nonbelief in, 286, 287, 288–93

Goldsworthy, Andrew, 114–16

Gordon, Linda, 72n15

Gordon, Richard, 38, 73n20

*Gospel of Ramakrishna, The,* xii

Gould, Stephen Jay, 68n3

Great Mother worlds, 272

Great Pyrenees (dogs), 155, 157–71; breeder of, 127–31; as dual-purpose dog, 164–65; emerging approaches to predator control in western United States and, 162–66; history of breed, 160–62; region of origin, 157; reintroduction of gray wolf and, 168–69

Great Pyrenees Club of America, 164; National Specialty Show, 163, 164

Great Pyrenees Discussion List, 128, 164

Great Pyrenees Standard, 162

Green, Jeffrey, 164, 166

Greenham Women's Peace Camp, 13

Green Revolution technologies, 42, 73n21

Gregory, Judith, 73n20

Griffin, Susan, 53, 75n30

Grossman, Rachel, 36–37, 73n119

Gubar, Susan, 75n32

Haas, Violet, 74n25

habitat regeneration, 234–35

Hacker, Sally, 74n25

Hampshire College, New England Farm Center, 166–67

happiness: animal, 143–45; practice of joy, 128, 129, 152, 154, 227, 244, 252–56

Haraway, Donna J., 73n17, 74nn23–25; attraction to "word made flesh" trope, xiii; conversation between Wolfe, Hogness, and, 201–96; doing biopolitics, xii; feminist lineages, 262–67; in History of Consciousness Department, UC at Santa Cruz, vii; on original context of "Cyborg Manifesto," 202–7; writing style of, viii–ix. *See also* Catholic background, influence of; "word made flesh," trope of

Harding, Marco, 130, 131–34, 153, 230

Harding, Sandra, 68n3, 72n14

Harding, Susan, 203, 267; on

global warming/climate change complex, 287

*Harper's* magazine, 143

Harper Trois Fontaine, Mme. Jeanne, 160–61

Hartmann, Heidi, 73n20

Hartsock, Nancy, 72n14, 203

Hawaii: Haraway's experiences in, 203–4, 205

Hearne, Vicki, x, 138–48, 152, 153, 225, 244; on companion "animal happiness," 143–45; dog training method, 138–45; opposition to discourse of animal rights movement, 138–39, 142–45

Heidegger, Martin, 292

Henifin, Mary Sue, 68n3

herders (dogs), 155; Australian Shepherds, 94, 155, 172–79; livestock guardian dogs compared to, 158–60; strong prey drive of, 159

Hesiod, 273

heteroglossia, cyborg, 68, 70n5, 75n29

high technology: challenging Western traditional dualisms in science fiction, 60–61;

*Index*

dualisms in social practices, symbolic formulations, and physical artifacts associated with, 14–15; high-tech military, 42, 75n33. *See also* technologies, new

High Technology Professionals for Peace, 74n26

Hintikka, Merrill, 72n14

"History of Agility" (Fender), 149

*History of Dogs in the Early Americas, A* (Schwartz), 105

*History of Sexuality* (Foucault), 69n5

Hoffman, Margaret, 163

Hogness, Eric Rusten, 32, 201, 204, 261

holism, 280–81

holocaust, 232; equating of killing of Jews in Nazi Germany with animal industrial complex, 142; of nonhuman life, 229–30, 231, 232

home: impact of social relations mediated and enforced by new technologies on, 46

Homebound Hounds Program, 186

homework economy, 37–45, 73nn20–21; development of new time arrangements to facilitate, 47; family of, 41; intensification of demands on women to sustain daily life for family in, 39–40; projections for worldwide structural unemployment stemming from new technologies in, 41–42; work redefined as both literally female and feminized in, 38–45; as world capitalist organizational structure, 38–45, 73n21

hooks, bell, 71n11

*"Horses, Hounds, and Jeffersonian Happiness: What's Wrong with Animal Rights?"* (Hearne), 143

*HOW(ever)*, 70n5

Hrdy, Sarah Blaffer, 74n23

Hubbard, Ruth, 68n3

Hull, Gloria, 71n11

human and animal: boundary between, 10–11

humanism: humanist technophiliac narcissism, neurosis of, 125; Marxian, 23; Western humanist sense of origin story, 8

humanities, 260, 261–62

humility: negative way as serious injunction to, 279–80

hunger: new technologies' effect on, 42. *See also* food production

Hustak, Carla, 253

Hutchinson, G. Evelyn, 241, 264

idealism: dialogue between materialism and, 11

identity(ies): Chicana constructions of, Malinche story and, 56–57; fractured, 16–27; limits of identification, 19–21; monsters and establishment of modern, 65; myth about political, 52–68; negative, 18; sources of crisis in political, 17

immunobiology, 34, 35; apparatus of immunology in microbiology, 249; immune systems as part of naturecultures, 94, 122, 220, 245–46

Impy (dog), 163

indigenous traditions: writings of, 284

informal markets, 47

informatics of domination, 28–37; control strategies in, 31–32; integration/exploitation of women into, 33; as massive intensification of insecurity and cultural impoverishment, 49–50; race in, 31; sexual reproduction in, 30–31; transitions from organics of domination to, chart of, 28–30, 72n17; women's "place" in integrated circuit, vision of, 46–49

innocence: cyborgs' lack of, 9, 65; cyborg writing about power to survive not based on original, 55–58, 62, 65; living in noninnocence, 235–36; noninnocence of category "woman," 21; Western epistemological imperatives to construct revolutionary subject from position of, 58

integrated circuit: women in, 36–37, 45–52

International Fund for Agricultural Development, 74n21

interpellation, 108–9

intersubjectivity: meaning of, 133; seeking to inhabit intersubjective world, 125; significant otherness and, 133–34

*In the Company of Animals* (Serpell), 105–6

invasive species, 235, 236

Irigaray, Luce, 53, 75n29

irony, viii, 5, 75n33, 221; in

"Companion Species Manifesto," 167, 221; of cyborg's lack of origin story, 8; of MacKinnon's "ontology" constructing a nonsubject, 24; refiguring, in "Cyborg Manifesto," 20–21, 52, 65, 96, 208–12, 221

ISIS (Women's International Information and Communication Service), 68n4

island ecosystem: invasive species in, 235; noninnocence of killing to preserve, 236

Jaggar, Alison, 71n13

James, William, 290, 291

Jameson, Frederic, 40, 70n7, 210; on postmodernism, 69n5

Jefferson, Thomas, 144; ideas of property and happiness remodeled by Hearne, 144–45

Jewishness: status of, 230

Jews: Holocaust and, 142

Johns Hopkins: teaching at, 203

joy, practice of, 252–56; play and, 227, 244, 253; of sharing life with different being, 128, 129, 152, 154, 227, 244. *See also* happiness

Kac, Eduardo, 222

Kahn, Douglas, 75n31

*Kainos*, 294, 296

Kamin, Leon J., 68n3

Katrina, Hurricane: "refugees" from, 252

Keller, Evelyn Fox, 68n3, 74n25

kennel club breeds of LGDs, 157–58

Keno (dog), 176

killable, making, 233, 234, 235–36, 271

Kimball, Linda, 73n20

kin, making, 223–26, 296; kinship-making apparatus and adoption of dog, 186–87; practice of joy and, 252–56

*Kindred* (Butler), 62

King, Katie, 19–20, 54, 61, 71n13, 74n28, 154

Kingston, Maxine Hong, 75n32

"kitchen-sink syndrome," 292

Klein, Hilary, 8, 68, 68n2

Knorr-Cetina, Karin, 69n4

knowledge: historical situation in ways of knowing, 205–6; negative way of, 141, 265; working toward unknowing, 212–14

Koehler, William R., 139

Komondor (guardian dog), 164, 167

Kramarae, Cheris, 75n31

Kristeva, Julia, 26, 72n16

Kuhn, Thomas, 69n4

Kuvasz dogs, 157

labor: experience as living histori-cal, 155; homework economy, 37–45, 73nn20–21; impact of social relations mediated and enforced by new technologies on, 47–48; international divi-sion of, 37, 73n21; in MacKin-non's radical feminism, 24–25; in Marxian socialism, 22, 23; microelectronics mediating translations of, 36; ontological structure of, unity of women and epistemology based on, 23; power of new communications technologies to integrate and control, 39; reformulation of expectations, culture, work, and reproduction for large sci-entific and technical workforce, 44; in socialist-feminism, 22–23; technical restructuring of labor processes, 50

"Lacklein" (Sofoulis), 8

Lacy, William, 74n21

LAG (Livermore Action Group), 15–16

landscape–human couplings, 114–16

language: cyborg heteroglossia, 68, 70n5, 75n29; feminist dream of common, 52; Hearne's dog training method using ordi-nary, 140; metaplasm, 112–13, 145, 193; need for other nouns and pronouns for kin genres of companion species, 187; sto-ries about, in writing by U.S. women of color, 56–57; tech-nobabble, 69n5; translation of world into problem of coding, 34–35

Latour, Bruno, 37, 69n4, 154, 268, 281, 284, 288, 289–93; feminist thinking and, 291–92; Stengers and, 290–91

Leffler, Ann, 151

LeGuin, Ursula, 282–83, 296

Lerner, Gerda, 75n32

Levidow, Les, 68n4

Lévi-Strauss, Claude, 75n31

Lewontin, R. C., 68n3

*Index*

319

*Index*

*Index*

redefining pleasures and politics of embodiment and feminist writing, 61–64

Moore, Jason, 239

Moraga, Cherríe, 56–57, 75n32

Morey, Darcy, 121

Morgan, Robin, 75n32

Mothers and Others Day action at nuclear weapons testing facility in Nevada (1987), 70n9

Mowatt, Twig, 180

Mulkay, Michael, 69n4

multidirectional gene flow, 101

multinational corporations: language of, 69n5; Third World women as preferred labor force for science-based, 38, 54

mutual adaptation as coevolution, 122

*My Dog Tulip* (Ackerley), 125–26

Myers, Natasha, 253

naming: negative way of. *See* negative way of naming

narcissism: caninophiliac, 124; humanist technophiliac, 125

Nash, June, 73n19

Nash, Roderick, 74n24

National Defense Education Act, 203

National Science Foundation, 74n25

Native Americans, 105, 174, 189

nature: construction of popular meanings of, 74n24; textualization in postmodernist theory undermining certainty of, 12

naturecultures, emergent, 93–117, 221; art of, 114–16; companions, 103–6; companion species, 107–17; example of, 143; immune systems as part of, 94, 122, 220, 245–46; implosion of nature and culture, 108–11; prehensions, 98–102

*Nausicaä of the Valley of the Wind* (Miyazaki), 296

Navajo sheep herding practices, 165, 167

necropolitics, 229

negative identities, 18

"negative" knowing: love as, 141, 265

negative way of naming, 265, 274, 276–85; in finitude, 278–79; as mode of thought originally developed in relation to question of God, 277–78

network ideological image of women's place in integrated circuit, 45–52; idealized social locations in, 46–49

Neumaier, Diane, 75n31

*Never Alone* (game), 296

"New Industrial Revolution," 37

New Orleans: Hurricane Katrina and "refugees" from, 252

*New York Times,* 73n21, 168

Nicholas of Cusa, 279

*Nop's Hope* (McCaig), 129

*Nop's Trial* (McCaig), 129

normalization, 69n5

Noske, Barbara, 126, 217, 232, 264

"Notes of a Sports Writer's Daughter" (Haraway), xi, 190–93; emergent naturecultures, 93–117; reporting the facts as rule of, 109–11; training stories from, 132–54

nothing-but-critique approach: refusal of, 211

nuclear weapons testing facility in Nevada, 1987 Mothers and Others Day action at, 70n9

Nussbaum, Karen, 73n20

objectification, sexual, 24–25

O'Brien, Mary, 72n14

*One-Dimensional Man* (Marcuse), 14

Ong, Aihwa (87), 73n19, 76n34

Ong, Walter (82), 75n31

ontological choreographies, 193, 224; beauty of, in skilled human–dog interaction, 143; in technoscience, 100, 103

ontology(ies): cyborg as our, 7–8; emergent, 99; grounding "Western" epistemology, loss of, 12; using different ontological approaches, 289

oppositional consciousness, 17–19

organics of domination, 28–30; cannibalization of dualisms of, 32; transitions to informatics of domination, 28–37, 72n17

orientalism, 71n12; deconstruction of, 19

original unity myth, 8

origin story(ies): by cyborg authors, 55–56; cyborgs' lack of, 8; evolution stories of dogs, 117–24; in feminist science fiction, 62; phallogocentric, 55–56, 57; in "Western," humanist sense, 8

*Os Mil Nomes de Gaia* / The Thousand Names of Gaia (meeting, Brazil), 273–74

*Index*

323

otherness-in-connection: alertness to, 142. *See also* significant otherness

*ouroboros* (snake-swallowing-its-own-tail), 277, 280–81

overhygienization, 248

Oxford, Gayle, 178

Oxford, Shannon, 178

paganism, feminist, 53

paid workplace: women's "place" in, 47–48

Park, Katherine, 76n36

partial connections, relations of, 100, 117, 140, 262; legacy inherited through companion species, 188–90. *See also* significant otherness

patriarchal nuclear family, 41

Pedigree Pet Foods, 149

Peirce, Charles, 268, 291

Perloff, Marjorie, 70n5

Perucci, Carolyn, 74n25

pet medicine, 256–60

pet relationships, 128–29. *See also* "Companion Species Manifesto, The" (Haraway); dogs

Petschesky, Rosalind, 42

Pfafflin, Sheila M., 73n19

phallogocentrism, 55–56, 57

pharmakon: logic of, 220–21

phenylpropanolamine (PPA), 257

physical and nonphysical: breach of boundary between, 12–14

Piercy, Marge, 204

Piven, Frances Fox, 73n20

plantation system, 239

plasticities, ecological developmental, 123

Plato, 70n7

play: game story, 149–54; as general topic for further investigation, 225–26; human–animal, 93–94, 103, 129, 136, 138, 146, 149–54, 189–90, 216–17; irony and, 5; practice of joy and, 227, 244, 253; sexual, between dog playmates, 191–93; of writing and reading the world, 11–12, 31, 55, 59

Plumwood, Val, 256, 264

pluralist model, democratic liberal, 286

police dog training: agility training's origins in, 148

political identity: crisis in, 17; limits of, 19–20; myth of, 52–68; oppositional consciousness as model of, 17–19

*Index*

324

*Index*

*Index*

oppositional consciousness, 17–19

sanitation: questions of, 246–48

Santa Rita Jail, 14

Satos (dogs), 180–89, 251–52

Save-a-Sato Foundation, 181, 183–85

scale-making, 155–56

Schiebinger, Londa, 74n25

school: impact of social relations mediated and enforced by new technologies on, 48–49

Schwartz, Marion, 105

science and technology, social relations of, 37, 43–44; reconstruction socialist-feminist politics through theory and practice addressed to, 33–37, 44, 50; structural rearrangements related to, 50–51; taking responsibility for, 67; women's historical locations in advanced industrial societies restructured partly through, 45–52

science fiction: boundary between social reality and, 6; cyborgs in, 6, 9–10, 54, 61–68; feminist, 54–68, 74n28; feminist, boundary breakdowns in, 63–64; feminist, constructions of monstrous selves in, 54, 61–68, 76n36; feminist, constructions of women of color in, 54–62; high-tech culture challenging Western traditional dualisms in, 60–61; intersection of feminist theory and colonial discourse in, 64

Science for the People, 68n3

Science magazine: article on dog evolution, 118

Science Policy Research Unit, 73n19

Sciences and Fiction conference, 201

Sciences Po in Paris, 293

scientific culture: boundary between human and animal, breach of, 10–11; communications technologies and biotechnologies, 33–36, 39, 41–43; consumption of scientific origin stories, 118–24; dualisms in social practices, symbolic formulations, and physical artifacts associated with, 14–15; feminist science, possibility of, 44–45, 52; growing up in, 110. See also science and technology, social relations of

Scott, Patricia Bell, 71n11

Scott, Ridley, 60

sculptures: as companion species, 114–16

secular Catholic: Haraway as, 267. *See also* Catholic background, influence of

secular Protestants, 267, 268

secular separatists, 267

SEIU's District 925, 50

self: cyborg as disassembled and reassembled, postmodern collective and personal, 33; as the One who is not dominated, 60

semiotics, 291; of Catholicism (word made flesh), xii–xiii, 107, 110, 266, 268–69, 274–75, 279, 280, 292; material, 268–75, 280, 281

Senac-Lagrange, Bernard, 160

separate spheres: white bourgeois ideology of, 41

separation: of church and state, 267; of the signifier and the signified, 268

Serpell, James A., 105

Service Employees International Union, 74n27

Seven Sisters, 71n11

sex: cyborg, xiii, 6; nonreproductive, 93, 94, 191–93, 220, 224, 225, 255; sexual reproduction in new informatics of domination, 30–32, 49

"Sex, Mind, and Profit" (Haraway), 250

sexual appropriation in MacKinnon's radical feminism, 24–25

sexuality: social relations of reproduction and, 43–44

sexual objectification, 24–25

Sharp, C. A., 177–79

sheep herding: Australian Shepherds and, 173–76, 177

shelter dogs, 184–86; adopting, 185–87

*Ship Who Sang, The* (McCaffrey), 61

Siegel, Lenny, 74n26

significant otherness, 101, 120, 228; alertness to demands of, 152; intersubjectivity and, 133–34; joint lives of dogs and people bonded in, 108, 116; relations of, 100, 116–17, 172; seeking category of one's own in, 188–89; significant otherness-in-connection, Hearne's dog training method and, 142

*Index*

329

*Signs: Journal of Women in Culture and Society,* 75n29

*Silence of the Lambs, The* (film): Wolfe and Elmer on, 269–70

Silicon Valley: alliances across the scientific-technical hierarchies, 45; women's employment in, 38

simulacra: microelectronics as technical basis of, 36; Plato's definition of the simulacrum, 70n7

Sisler, Jay, 176–77

*Sister Outsider* (Lorde), 54, 56–57

"Situated Knowledges" (Haraway), 250

Sklar, Holly, 73n20

Smith, Barbara, 71nn10–11

Smith, Dorothy, 72n14

Smuts, Barbara, 126

Sno-Bear (dog), 163

socialism, Marxian, 22

socialist-feminism, 27, 72n15; basic analytic strategies of Marxism adopted by, 22–23; political myth for, 21; reconstructing socialist-feminist politics through theory and practice addressed to social relations of science and technology, 33–37, 44, 50; repro-duction, socialist-feminist sense of, 22–23, 25–26; silence about race, 26

*Socialist Review,* 72n15, 211; West Coast Collective, 202

social reality, 5–6; boundary between science fiction and, 6

social relations: of both sexuality and of reproduction, 43; of new technologies, 43–44, 46–49; of science and technology. *See* science and technology, social relations of

sociobiological evolutionary theory, 34; origin stories, 43, 74n23

Sofia [Sofoulis], Zoë, 7–8, 15, 32, 64, 68n1

Sontag, Susan, 74n24

species, 107–17; debates about category "species," 106–7; destructive, Anthropocene and, 237–43; distinctions, extraction of value based on, 260; four tones resonating in term, 107–8; invasive, 235, 236; multiple meanings, 216; race and, xi, 260–61; recovery plans for, 234–35. *See also* "Companion Species Manifesto, The" (Haraway)

*Index*

systems theory (cybernetics), vii, 34, 263–64; later-generation, 279

Tadiar, Neferti, 155
talent, companion animal, 143
*Tales of Nevèrÿon* (Delany), 62
Taylorism, 6
technological determinism, 11–12
technologies, new: biotechnologies, 33–36, 121; convention of ideologically taming militarized high technology, 75n33; Green Revolution, 42, 73n21; heightened sense of connection to our tools, 60–61; high-tech culture challenging Western traditional dualisms, 60–61; homework economy and, 38–45; militarized high technology, 42, 75n33; privatization and, 42–43, 48; social relations of, 43–44; social relations of, impact on idealized social locations in integrated circuit, 46–49. *See also* communications technologies and biotechnologies
textualization in poststructuralist, postmodernist theory, 12

thanatopolitics, 226
*Theogony* (Hesiod), 273
Third World: coalition politics emerging from, 71n10; pressures on land in, 73n21; teenage women in industrializing areas of, as sole or major source of cash wage, 40; women, as preferred labor force for science-based multinationals, 38, 54
Thompson, Charis, 100–101, 224
Tinbergen, Niko, 166
Tiptree, James, Jr., 52, 62; works of, 75n28
totalizing theory, 26; MacKinnon's radical feminism as caricature of, 23–24; as mistake, 52, 67
tourism industry, 43
training stories about dogs, 132–54, 154; agility training, 146–54; game story, 149–54; harsh beauty, 138–45; positive bondage, 134–38
Traweek, Sharon, 68n4
Treichler, Paula, 35, 75n31
Trinh T. Minh-ha, 71n10

*Index*

332

tropes, 123; defined, 112. *See also* "word made flesh," trope of

Tsing, Anna, 155–56, 201, 247–48, 295

unconditional love, 124, 126, 129

unemployment, worldwide structural, 41–42

United States Australian Shepherd Association (USASA), 177

United States Dog Agility Association (USDAA), 149

unity: through affinity, 17, 18, 19–20; fractured identities and, 16–27; grounds for hope in emerging bases for new kinds of, 51–52; through incorporation, 17–19; need and opportunity for, 21; of people trying to resist worldwide intensification of domination, need for, 15

University of California at Santa Cruz, 201; History of Consciousness program, 209, 229

University of California–Davis, 162, 163

University of Hawaii: teaching at, 203, 204–5

U.S. Agency for International Development, 73n21

U.S. Department of Agriculture, 162, 163, 169; project for use of livestock guardian dogs, 165;

U.S. Sheep Experiment Station, 164

U.S. Department of the Interior, 168–69

van Dooren, Thom, 231, 256

Varley, John, 52, 62

Verran, Helen, 99, 189

*Vibrio* bacteria: light-sensing organs of *Euprymma scolopes* and, 123

video games culture, 42–43

Vietnam War: McNamara plan, 205

Vilá, Carles, 120

violence: family, 72n15; making killable, 233, 234, 235–36, 271; of making live, 229, 231

"Virtual Speculum" (Haraway), 250

visualization: bodily boundaries newly permeable to, 43–44; technologies of, 44

Viveiros de Castro, Eduardo, 273

von Uexküll, Jakob, 209–10

*Index*

*Index*

ily, 39–40; "New Industrial Revolution" and homework economy, 37–45, 73nn20–21; subsistence food production by, 42, 73n21; Third World, as preferred labor force for science-based multinationals, 38, 54; work redefined as both literally female and feminized, 38–45
*Women and Poverty,* 73n20
women of color: constructions in feminist science fiction of, 54–62; definition of, 18; literacy as special mark of, 54–55; oppositional consciousness model of political identity and, 17–19; poetry and stories about writing by, 55–59, 75n32; as preferred labor force for the science-based industries, 38, 54; understood as cyborg identity, 54
women-of-color feminism, 207
women's culture, 19–20
women's health movement: exposure of DES scandal by, 258
Women's International Information and Communication Service (ISIS), 68n4
women's movement: feminist

taxonomization of, 19–20; politics of race and culture in United States, 20. *See also* feminism
Woodruff, Robert, 164, 166
Woolf, Virginia, 179
Woolgar, Steve, 69n4
word as fetish, 274–75, 277, 292
*Word for World Is Forest, The* (LeGuin), 282
"word made flesh," trope of, xii–xiii, 110, 266, 279; doctrine of the Real Presence and, 107; material semiotics of Catholicism and, 268–69, 280; word as fetish and, 274–75, 292
work: developing forms of collective struggle for women in paid, 50; impact of social relations mediated and enforced by new technologies on paid workplace, 47–48; redefined as both literally female and feminized, homework economy and, 38–45. *See also* labor
working animals, 230; herding dogs, 94, 155, 172–79; livestock guardian dogs, 155, 157–71
working class, 37, 47, 50
World Food Day (1984), 73n21

## *Index*

*Index*

CARY WOLFE, SERIES EDITOR

*Donna J. Haraway* is Distinguished Professor Emerita in the History of Consciousness Department at the University of California at Santa Cruz, where she is also an affiliated faculty member in anthropology, feminist studies, environmental studies, and film and digital media. Her books include *When Species Meet* (Minnesota, 2008); *Crystals, Fabrics, and Fields*; *The Haraway Reader*; *Modest_Witness@Second_Millennium*; *Simians, Cyborgs, and Women: The Reinvention of Nature*; and *Primate Visions: Gender, Race, and Nature in the World of Modern Science.*

*Cary Wolfe* is Bruce and Elizabeth Dunlevie Professor of English at Rice University, where he is founding director of 3CT: The Center for Critical and Cultural Theory. His books include *Animal Rites: American Culture, the Discourse of Species, and Posthumanist Theory*; *What Is Posthumanism?* (Minnesota, 2010); and *Before the Law: Humans and Other Animals in a Biopolitical Frame.* His edited and coedited collections include *Zoontologies: The Question of the Animal* (Minnesota, 2003) and *The Other Emerson* (Minnesota, 2010).